WEYERHAEUSER ENVIRONMENTAL BOOKS
William Cronon, Editor

Weyerhaeuser Environmental Books explore human relationships with natural environments in all their variety and complexity. They seek to cast new light on the ways that natural systems affect human communities, the ways that people affect the environments of which they are a part, and the ways that different cultural conceptions of nature profoundly shape our sense of the world around us.

WEYERHAEUSER ENVIRONMENTAL BOOKS

The Natural History of Puget Sound Country by Arthur R. Kruckeberg

Forest Dreams, Forest Nightmares: The Paradox of Old Growth in the Inland West by Nancy Langston

Landscapes of Promise: The Oregon Story, 1800–1940 by William G. Robbins

The Dawn of Conservation Diplomacy: Canadian-American Wildlife Protection Treaties in the Progressive Era by Kurkpatrick Dorsey

Irrigated Eden: The Making of an Agricultural Landscape in the American West by Mark Fiege

Making Salmon: An Environmental History of the Northwest Fisheries Crisis by Joseph E. Taylor III

George Perkins Marsh, Prophet of Conservation by David Lowenthal

Driven Wild: How the Fight against Automobiles Launched the Modern Wilderness Movement by Paul S. Sutter

The Rhine: An Eco-Biography, 1815–2000 by Mark Cioc

Where Land and Water Meet: A Western Landscape Transformed by Nancy Langston

WEYERHAEUSER ENVIRONMENTAL CLASSICS

The Great Columbia Plain: A Historical Geography, 1805–1910 by D. W. Meinig

Mountain Gloom and Mountain Glory: The Development of the Aesthetics of the Infinite by Marjorie Hope Nicolson

Tutira: The Story of a New Zealand Sheep Station by H. Guthrie-Smith

A Symbol of Wilderness: Echo Park and the American Conservation Movement by Mark W. T. Harvey

Man and Nature: Or, Physical Geography as Modified by Human Action by George Perkins Marsh; edited and annotated by David Lowenthal

CYCLE OF FIRE BY STEPHEN J. PYNE

Fire: A Brief History

World Fire: The Culture of Fire on Earth

Vestal Fire: An Environmental History, Told through Fire, of Europe and Europe's Encounter with the World

Fire in America: A Cultural History of Wildland and Rural Fire

Burning Bush: A Fire History of Australia

The Ice: A Journey to Antarctica

Where Land & Water Meet

A WESTERN LANDSCAPE TRANSFORMED

Nancy Langston

FOREWORD BY WILLIAM CRONON

UNIVERSITY OF WASHINGTON PRESS

Seattle and London

Where Land and Water Meet: A Western Landscape Transformed has
been published with the assistance of a grant from the Weyerhaeuser
Environmental Books Endowment, established by the Weyerhaeuser
Company Foundation, members of the Weyerhaeuser family,
and Janet and John Creighton.

Library of Congress Cataloging-in-Publication Data
Langston, Nancy.
Where land and water meet : a Western landscape transformed / Nancy Langston.
      p.   cm.—(Weyerhaeuser environmental book)
Includes bibliographical references (p.).
ISBN 0-295-98307-8
1. Wetland management—Oregon—Malheur National Wildlife Refuge.
2. Nature—Effect of human beings on—Oregon—Malheur National Wildlife Refuge.
3. Malheur National Wildlife Refuge (Or.)—History
I. Title.   II. Series.
QH76.5.O7L36 2003   333.91'8'09795—DC21   2002035784

CONTENTS

FOREWORD / On the Margins

William Cronon

Few subjects have attracted more attention from scholars of western history than water. Indeed, many historians and geographers have gone so far as to argue that water—or rather, the lack thereof—is the characteristic that most defines the trans-Mississippi West. Books and articles on western water are so numerous that entire volumes have been devoted simply to cataloguing them. But this vast literature nonetheless has some surprising lacunae. More than anything else, scholars have focused on the struggle to transport water from a handful of western rivers to the cities and agricultural districts that could not grow without it. We have dozens of studies about how the Colorado, the Columbia, the Owens Valley, and Hetch Hetchy came to be dammed, and how Seattle, San Francisco, and most of all Los Angeles flourished as a result. Our understanding of western water law and policy has few rivals in the field. On the other hand, we know much less about how all this water eventually came to be consumed. How did western lifestyles become so dependent in such profligate ways on so scarce a resource? Surprisingly little has been written about the cultural history of water: people's conceptions of what it means, how it should be used, and why its control is such a powerful symbol of progress. Perhaps most intriguingly, we know least of all about those places in the American West where water was most abundant: the wetlands.

Why should this be so? It's probably fair to say that wetlands have attracted less attention than they deserve not just in the arid regions of the trans-Mississippi West but even in landscapes such as the Middle West or the American South, where they constitute a much larger portion of the whole. With a few honorable exceptions—stretching from Ann Vileisis's *Discovering the Unknown Landscape: A History of America's Wetlands* and Hugh Prince's *Wetlands of the American Midwest* all the way back to Marjory Stoneman Douglas's classic *The Everglades: River of Grass*—the paucity of book-length studies of marshes, swamps, bogs, fens, bayous, sloughs, mires, muskegs, and morasses would seem to suggest that the most interesting things about such places are the peculiar names we assign to them.

In fact, scholarly indifference in this case reflects a much more encompassing cultural blind spot. Throughout most of American history, wetlands were "wastes": useless, worthless, bothersome places that blocked travel, bred mosquitoes, frustrated settlement, and generally threw up the most annoying and inconvenient barriers to human progress. One tried to avoid them if one could, cursed one's misfortune if one could not, and generally looked forward to the day when they could be drained and turned toward more productive ends. Up until the second half of the twentieth century, the most common response to American wetlands was almost always to drain them, and yet we have had surprisingly few historical studies of this crucial landscape-transforming process.

It is just this gap that *Troubled Waters* seeks to fill. Its author, Nancy Langston, is among the few environmental historians writing today who is trained equally well in ecology and history, so that she is as comfortable in the field as in the archive. She has a special genius for identifying relatively little-known places that can nonetheless serve as paradigmatic examples of ecological and historical processes whose implications extend far beyond the boundaries of her particular study. In *Forest Dreams, Forest Nightmares*, she demonstrated with a remarkable depth of insight how the Blue Mountains in northeastern Oregon could stand for an entire century of Forest Service management (and mismanagement) of forests in the arid West. Now, in this book, she performs the same feat for the lands of Malheur National Wildlife Refuge in the Blitzen River watershed of southeastern Oregon—a wetland the size of Massachusetts, Connecticut, and Rhode Island combined. Although Malheur is hardly one of the West's top destinations for tourists, most of whom have never even heard of it, the same can hardly be said for ducks. Indeed, it is no exaggeration to say that Malheur plays an absolutely crucial role in the survival of migratory birds along the entire West Coast, making it easily one of the most important wildlife refuges in both the United States and Canada.

The story Langston tells about this vast wetland has all the complexities, contradictions, and surprising insights that have become her trademarks as a scholar. Her chapters trace a progression of human inhabitants who have left their marks on the region, beginning with the native people for whom it was an indispensable source of plants to gather and animals to hunt. By the second half of the nineteenth century, these Indian land uses were supplemented, if not replaced, by vast cattle-grazing operations. Although modern environmentalists often view ranchers and their herds as unrelentingly destructive of the lands they graze, Langston argues that these early cattle operations in the Malheur Lake Basin were surprisingly sensitive in their use of the environment. The real damage began to be done at the turn of the century as homesteaders and irrigation engineers promoted a much more aggressive vision of wetlands drained in the service of human progress. By defending their work as a compelling contribution to democracy and social justice, they redefined the rich wetland ecosystem into a place of no value: a waste. And so the Malheur Lake Basin, like so many other marshes and swamps across the nation, began to be channeled, tiled, dredged, and drained so that the old republican dream of America as a nation of Jeffersonian yeoman farmers could persist into the twentieth century.

However compelling this reclamationist dream might seem to farmers and engineers, it was no friend to ducks. By the early decades of the twentieth century, waterfowl populations all along the Pacific Flyway were in serious decline from over-hunting and loss of habitat. Worse, the extensive manipulation of the wetland ecosystem had not fulfilled its human promise either: farmers and ranchers alike were going bankrupt not just because of the economic crisis of the 1920s and 1930s, but because the environmental transformations they had brought about had undermined their own prosperity. By 1934, conditions had become so bad that the federal government began purchasing the failed farms and ranches to create Malheur National Wildlife Refuge.

Refuge managers now faced precisely the opposite challenge from that of their predecessors: reversing nearly half a century of drainage to reflood the land and restore the health of the wetland ecosystem. This was a new kind of "reclamation," effectively turning the old definition on its head. They nonetheless pursued it with much the same aggressive energy, even many of the same tools, as the irrigationists who had preceded them: building dams, dredging ponds, rerouting watercourses, and digging hundreds of miles of new canals to create a wholly reengineered landscape so that water could be carefully routed to places where it was needed most. What they created was not a restored wilderness—that was never their intention—but rather a vast system committed to the production of avian flesh. Whereas ranchers had managed the land to maximize the production of cattle, and farmers to maximize the production of alfalfa, the refuge managers now sought to maximize their yield of ducks. And, as so often happens when people pursue their goals with such single-minded intensity, not everything turned out quite as the managers expected. By the late twentieth century, wildlife managers were facing a host of challenges, human and ecological alike, that were requiring them to rethink the goals that had seemed so straightforward when Malheur National Wildlife Refuge was first created in the 1930s.

The story of the Malheur Lake Basin is both intricate and fascinating, and I will not try to summarize all of its twists and turns in this brief foreword. But if I wished to distil the lessons Nancy Langston draws from this complex tale, I might say the following. Human single-mindedness can tempt even the most enlightened managers into focusing their attention too narrowly on what most interests them. Great things can be achieved in this way, whether the goal is to produce human food, economic prosperity, or environmental protection. But often the things that interest us most—whether those things are cows or crops or ducks—are connected to other, less obvious things that we may not even notice until our failure to do so begins to cause trouble. When trouble happens, the trick is to pay attention in a new way. Doing so often means broadening our goals, managing for greater complexity, and—as Langston demonstrates so clearly in this fine book—revising our understanding of past actions that we thought we understood more clearly than in fact we did.

In the end, Nancy Langston offers two prescriptions for the ills that beset Malheur National Wildlife Refuge. One is "adaptive management": a recognition by conservationists and anyone else who must care for the health of a landscape that ecosystems are intrinsically dynamic and that no static management goals can ever protect them for long. Managing such systems for single organisms will almost never be a good idea. Furthermore, the complexity of an ecosystem is mirrored by the complexity of human communities and human politics, so that often the best way to protect ecosystems is to involve a wide range of human perspectives and human groups in the management process. The wisdom that results from their collective dialogue is likely to be more sound—and the resulting political decision-making more robust—than if a single group of managers monopolizes all decisions.

This in turn leads to Langston's second lesson. If adaptive management depends on a more dynamic understanding not just of natural ecosystems but of human communities, then by definition it must pay more attention to history. History, after all, is the study of dynamic systems, of change over time. It almost always reveals that things are more complicated than they seem. The way we understand the world today is rarely the way our predecessors understood it, and that will be no less true of our successors. Taking the long view can help protect us from our own time-boundedness, reminding us that goals pursued for one reason almost always have unforeseen consequences that will come back to haunt us if we define what we are doing too narrowly. If all of this is true, then the history Nancy Langston offers in *Where Land and Water Meet* has implications far beyond the Malheur Lake Basin.

ACKNOWLEDGMENTS

I am grateful to the many people who shared their love for eastern Oregon and their knowledge of riparian processes. The staff of Hart Mountain National Antelope Refuge and Malheur National Wildlife Refuge gave me every possible assistance. I am particularly indebted to Carla Burnside, Richard Roy, Gary Ivey, and Forrest Cameron for the many hours they spent talking to me, helping me find archival records and photographs, and, most important, taking me on field tours of Malheur National Wildlife Refuge. The thoughtful comments of Carla Burnside and Richard Roy on the manuscript helped me avoid many errors.

Many residents of Harney County talked with me, and their willingness to share their views made this a more balanced work. Fred Otley and his family were generous in spending time showing me riparian areas on their land and talking about ranching on Steens Mountain. Staff members of the Oregon Natural Desert Association, particularly Bill Marlett and the late Joy Belsky, were equally generous with their time.

Staff with the Bureau of Land Management were enormously helpful, particularly Wayne Elmore, who helped me understand riparian restoration, and Guy Sheeter, who took me on tours of restoration projects and spoke for hours about restoration work. Numerous staff members in the Hines office of the Bureau of Land Management, the headquarters of Malheur National Wildlife Refuge, the Harney County Historical Society, the Harney County Courthouse, the Oregon Historical Society, and the Harney County Library were most generous with their assistance. I thank them for enthusiastically helping with endless odd requests for files, data, photos, and information.

Over the years, provocative comments on various chapters and presented papers were provided by many people, including graduate students in the social forestry seminars and environmental history groups at the University of Wisconsin. In particular, Mark Fiege's comments on the manuscript were thorough, insightful, and generous. Bill Cronon's support and critical acumen improved this project enormously, and I am honored to work with him.

My thanks to the University of Washington Press staff, particularly Lorri Hagman, whose editorial work clarified my meaning time and time again, and Julidta Tarver, whose warmth and enthusiasm have made working with the Press a great pleasure. This project was supported by a generous fellowship from the National Humanities Center, and Center staff did everything possible to help with the research during my 1998 residency.

Without Frank Goodman's support, completing this book would have been a much harder task. I am very grateful for all he has done for me over the years I spent on this project.

Where Land and Water Meet

Introduction

"Can't keep the flood waters off. A world of water
goes over the surface of the land."

—Mr. P. G. Smith, rancher, 1918

One piece of land is never entirely separate from another, even if we think that a string of barbed wire forms an effective barrier between them. A bit of dirt kicked free by a cow finds its way into a stream and eventually gets deposited miles away. That sediment clogs the gills of a redband trout, far from where the cow had grazed. Moving water connects these places, weaving the threads of the landscape together. The places where water and land combine—the riparian zones—mediate these connections, and what happens in these zones affects areas far beyond their boundaries.

Federal resource managers have often managed riparian areas in ways that confound outsiders. Since the 1930s, for example, managers at Malheur National Wildlife Refuge in southeastern Oregon have restored riparian habitat by ditching wetlands, channelizing creeks, running cattle through riparian meadows, spraying herbicide over creeks, mowing down willows that managed to escape the poison, killing beaver that entered the system, and repeatedly pouring rotenone into the rivers and lakes. Given that riparian specialists now consider these activities to be among those most harmful to riparian health, one can't help but wonder: Why did managers do this? What led them to make such decisions? What were the effects—ecological and social—of such decisions? And what can we learn from their successes and their failures? A better understanding of the tangled history of riparian restoration can help formulate strategies for more effective resource management.

Along streams and lakes in the semiarid Great Basin, lush vegetation shadows the waters, forming cool riparian zones that are critical for life in dry places. Wetlands and streamside forests act as sponges, absorbing floodwaters, buffering nutrients, improving water quality, and serving as nurseries for many species. They help maintain water tables and keep streams flowing—critical functions in a dry land. Although they form less than 1 percent of the landscape, they provide habitat for at least 80 percent of terrestrial species—humans not least among these.[1] Where still water—ponds, lakes, seeps—meets the land, one finds what are technically termed lentic riparian zones, better known as wetlands. Where moving water—rivers, streams, creeks—meets the land, one finds lotic riparian zones. When people talk about riparian areas, they often are referring only to the latter type, along moving water. The two types have important structural and functional differences, but in this book I look at both, since I am most concerned with the ways people have transformed the boundaries between land and water.

Riparian areas—especially those in dry places—have suffered from human impacts more than most other landscapes. Nearly 90 percent of riparian areas on public lands across the West are now badly degraded, according to the General Accounting Office.[2] Why are the boundaries between water and land in such poor condition? To find answers, we must examine not just the ecological and hydrological processes at work. We must also pay attention to the human processes shaping riparian landscapes. Any ecosystem is a product of its history, and that history includes cultural as well as ecological forces. For thousands of years, people in the arid West concentrated their cultures and technologies in these boundary landscapes. Riparian areas are places of rich resource concentration—good places to fish, to drink, to plant a field of alfalfa, to mine for gold, to nibble at grasses if you happen to be a cow. They are often the only level places in sight, and so first tracks and trails, then railroads, and eventually interstate highways have been concentrated along riparian areas.

Mining, fur trapping, removal of beaver, dam construction, drainage, irrigation, road building, logging, grazing, and urbanization—all these have profoundly shaped riparian areas. Yet landscape change comes about not just because people graze cows, dam rivers, mine for gold, drain meadows, plow up river bottoms, and cut down trees but because they do these things in a world where nature, culture, ideas, and markets tangle together in complex ways. To restore damaged riparian areas, we need to understand not only the ecological and hydrological processes that have been disrupted but *why* those disruptions occurred.

Even though riparian areas received much of the brunt of Euro-American settlement in the West, for generations they were conceptually invisible. Not really water and not really land, they have fallen outside normal scientific categories of analysis. This has had major implications not just for research but also in the physical world of creeks, ducks, and cows. In this book, I offer a historical reading of the varied and contested meanings of riparian areas, focusing on the ways concern about them evolved as an issue over the past century. Although my study is relevant to riparian management throughout the arid regions of the United States, I focus my attention on a single case, the Malheur Lake Basin in southeastern Oregon (map 1). Home once to the largest cattle empires in the world, and then to one of the most important wildlife refuges in America, Malheur has long been an epicenter for explosive debates over effective land management.

When bird-watchers now flock to Malheur National Wildlife Refuge, they find what looks like a teeming wilderness along the Donner und Blitzen River, usually known as the Blitzen River (map 2). Riparian meadows, willow thickets, ponds, and marshes create an oasis in the high desert of the northern Great Basin used each year by up to twenty-five million birds. To the visitor, Malheur seems like a supremely wild refuge from both desert aridity and human industry. Yet wild as it seems, this landscape has been radically transformed by ranchers, irrigators, and wildlife managers. In chapter 1, I examine the ways nineteenth-century cattle barons managed this landscape for the increased production of cattle. Although early ranchers certainly simplified nature, they did so with surprising sensitivity to ecological conditions, as they sought to accommodate their grazing regimes to intermittent flooding. Yet, for all their ability to manage environmental complexity with some responsiveness, ranchers proved far more rigid in their approach to social and political complexity.

As chapter 2 argues, irrigation engineers and reclamationists joined with homesteaders in an uneasy alliance to disrupt the political hold of ranchers over the basin's ecological riches. In trying to achieve a progressive vision of social equity, reclamationists' actions led to drastic ecological changes, for their vision depended on engineering the movement of water across the land in ways far more radical than ranchers had ever attempted. Ironically, their progressive vision of social equity became hijacked by powerful ranchers who learned to shift their own identities—from cattle barons to irrigation developers—in their attempts to consolidate power in the basin.

By the early decades of the twentieth century, drainage and reclamation had led to a noticeable decline in ducks throughout the region and the nation, stimulating a national interest in conservation of waterfowl and their habitats. After four decades of overgrazing, irrigation withdrawals, grain agriculture, dredging, and channelization, followed by several years of drought, Malheur had become a dust bowl. Attempts to increase production by making wet lands drier and dry lands wetter had stripped the willows and cottonwoods from the banks, imprisoned the river in a ditch, and dried up the meadows and marshes. People did not fare any better than the land: ranches failed, livestock starved, homesteaders went bust, and the primary occupation in the valley became suing neighbors over water rights. Water control was an unmitigated disaster.

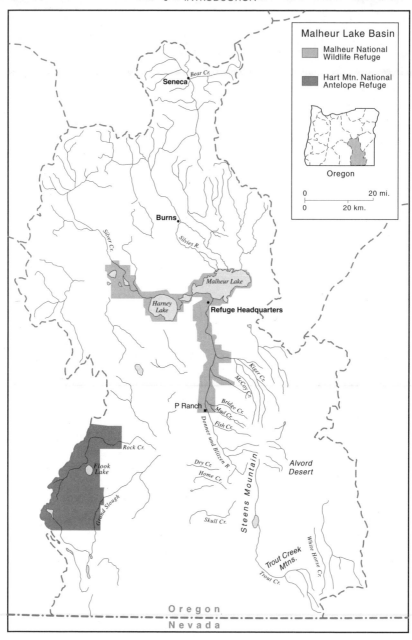

MAP 1. Malheur Lake Basin

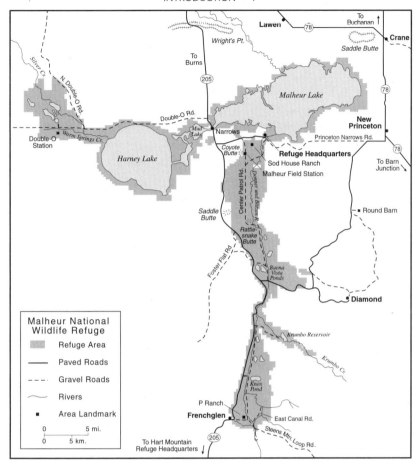

MAP 2. Malheur National Wildlife Refuge

Chapter 3 examines how conservationists won a major victory in 1934, when the failed cattle and irrigation empires along the Blitzen River were sold to the federal wildlife refuge system, beginning the expansion of an empire of ducks at Malheur. This event signaled the growing power of an urban elite's vision of conservation, a vision that was increasingly able to transform policies as well as landscapes. Yet although most conservationists were certainly part of an urban elite, their success came from the political alliances they formed with many of the most dispossessed people in the basin—the squatters and homesteaders who came to believe that the wildlife refuge managers would protect their interests better than would the ranchers or irrigators.

By the time Malheur Migratory Wildfowl Refuge was established, managers faced the daunting task of trying to restore an ecosystem that looked as if it had nearly been annihilated. Chapter 4 explores how, in response to what they saw as a crisis, refuge managers adopted drastic measures to reflood drained lands, reroute watercourses, and essentially manufacture new breeding areas for bird populations that seemed on the verge of extinction. With the help of men from the Civilian Conservation Corps (CCC), refuge staff bulldozed ponds, built dams, dug ditches, and extended hundreds of miles of canals along the valley. As one wildlife biologist on a neighboring refuge explained, whenever a refuge manager found some water in the desert, he tried to develop it—dam it, ditch it, impound it in a pond, or spread it out—anything but leave it alone.[3]

Managers built an empire of nature, a world aimed at increasing waterfowl production. Trying to maintain this empire led refuge staff into continued complications. When carp got into the lake and began making their way up the Blitzen River watershed, stirring up sediment and eliminating the sago pondweed that ducks required, refuge staff sprayed rotenone to kill the carp, along with anything else that was in the water, such as the native redband trout. When coyote and raven populations soared, lowering duck-nest success, refuge staff set out poisoned bait and then had to contend with rodent predation on eggs. When beaver returned to the valley and blocked up the irrigation ditches, staff trapped them out, even though the irrigation system was trying to replicate what beaver had created in the first place.[4] When willow began recuperating and spreading along the ditches and channelized banks of the Blitzen, staff mowed them down, since they were getting in the way of water management.

Today, when visitors to Malheur find out what kinds of intensive manipulations were needed to make this place seem wild, they are often dismayed. For many decades, refuge management was driven by visions of maximizing duck production, just as earlier management in the valley had been driven by hopes of maximizing cattle production. Such single-species management often led into blind alleys, where one technique created the need for a maze of corrections that in turn led to further complications. Yet managers' initial interventions seemed little short of miraculous, bringing the habitat for ducks back from the dead. After a few decades of federal management, the once desiccated riparian meadows along the Blitzen were again lush and green, water moved through the wetlands, and flocks of waterfowl returned to darken the skies.

The restored wetlands of Malheur Lake soon formed the centerpiece of a huge riparian marsh complex in southeastern Oregon—one large enough to cover Massachusetts, Connecticut, and Rhode Island. By 1987 refuge manager George Constantino could report that Malheur Refuge was "the most important refuge" along the Pacific Flyway "for water-oriented birds." In the oddly agricultural language of wildlife biologists, Malheur "produced" 84 percent of the West's great egrets, 55 percent of snowy egrets, and 68 percent of black-crowned night herons, representing "a major production area in the Pacific Flyway."[5] But for all its appearance of wildness, this had become an intensively managed landscape. Only in subsequent years have we learned of the subtle ways in which the aggressive management techniques of the early period have come back to haunt both the wetland and its managers. These transformations led the refuge staff into spirals of continual management complications and of escalating battles with ranchers, farmers, and now environmentalists. Chapter 5 examines the ways refuge policies have been transformed since the 1970s through floods, litigation, and new understandings of ecosystem complexity.

One of my core arguments in this book is about the need to recognize complexity. Integrating ecology and history allows us to make our understanding of environmental complexity much richer. But the moral of this story is not just that things are complex—I want to show the processes by which awareness of complexity can and has improved management. Conflict is central to these processes. For many people, conflict is a dirty word. Conflict has been a key part of American environmental politics, and many people—both environmentalists and critics of environmentalism—think that is a very bad thing. Yet conflicts among different users of Malheur Lake Basin eventually improved refuge management, for those conflicts disrupted the hold of narrow orthodoxies on resource management.

For generations, first ranchers and then refuge managers were able to gather enough power so that they did not need to acknowledge viewpoints other than their own. Ranchers moved away from their hold on the basin only when lawsuits, drought, and financial ruin in the 1930s forced changes in the balance of power. Four decades later, refuge managers were reluctant to modify their own ideologies until environmentalists used litigation against them. A set of escalating conflicts— conflicts that began as local issues and then became mediated by national institutions—eventually forced groups in Oregon to embrace a political process in which stakeholders coming from different perspectives had to jostle against, argue with, and listen to one another in ways that modified their actions and beliefs. Because no one can ever have perfect knowledge of how dynamic ecological systems work, this process moved the refuge toward much better solutions than any one group could have found on its own. Both environmental lawsuits and environmental change forced open a door through which new stories, new perspectives, and new assumptions could enter.

Before the 1930s, when ranchers could operate with absolute authority over what they thought was their own domain, they did not do a good job of protecting ecosystems. Likewise, by the 1970s, when refuge managers were permitted to operate with the authority of state power reinforcing their assumptions, they failed to adequately protect entire ecosystems. Only when political conflict in the 1970s forced them to allow other stakeholders to have a voice did they begin to question some of their own assumptions that seemed so self-evident when they hadn't been required to answer to anyone else.

As I argue in the last chapter, this is a vision of the promise of adaptive management—an iterative political and management process that yields new information about ecological and human systems, and then uses that information to develop policies that can respond to changing knowledge about a changing world. This final chapter places adaptive management within the context of American pragmatism, the philosophical tradition celebrating the democratic process and the scientific method as opposed to rationalist traditions that assert their claims to knowledge on the basis of absolutist and a priori reasoning. Pragmatism offers a useful framework for examining conflicts between the visions of ranchers, homesteaders, irrigationists, wildlife managers, and modern environmentalists, allowing us to understand what happened when their visions collided.

In the remote Malheur watershed, ranchers, irrigation speculators, farmers, and wildlife biologists competed for control of the uncertain boundaries between water and land. All of the groups that have lived and worked in the basin have changed the connections between water and land, and all of these changes have led to unintended consequences. But the moral of this story is not that everything people have done has degraded the ecosystem. The effects of different groups on the riparian landscapes were profoundly different. Some people tried to bring stability, hoping to create a predictable machine that could maximize agricultural or waterfowl production. They cut the connections between water and land, often with disastrous effects. In the short term, this increased cattle, hay, and birds. In the longer term, their efforts destabilized the riparian systems.

Yet not everyone made these mistakes. Some people, such as the earliest ranchers, manipulated water's boundary with land without ever hoping to achieve complete control over either, and without severing the connections between water and land. Like everyone else, they hoped to increase production of certain resources. But rather than separating water from the land, they adjusted their own practices to benefit from the movement of floodwaters across the riparian meadows. They did not try to create a simple machine out of nature, a machine utterly within the control of humans for producing ducks or cattle or crops.

My goal throughout this book is to tell the story of one small place that, for all its remoteness, has myriad ecological and political connections to much larger places. The history of this landscape reveals the many ways a landscape can become impoverished. Yet I also show that people can and have used that history to improve their relationships with the land. Surprisingly, perhaps, there are lots of success stories in this book—stories that suggest that although we certainly make plenty of mistakes, we can build on them to improve management, if we create a structure for adaptive management that allows for new stories to find a voice in the process.

1 / Ranchers in the Malheur Lake Basin

To trace water's journey across the Malheur Lake Basin, we should start where the Donner und Blitzen River begins: high on the west slopes of Steens Mountain, a tilted fault block that rears up into the skies nearly ten thousand feet above sea level. Because its massive flanks thrust five thousand feet above the surrounding Great Basin desert, Steens is an enormous snowmaker, and these snows feed the marshes below. Some of the snows melt slowly enough so that meltwater soaks into the mountain soil, seeping and creeping into springs that feed high-elevation streams protected by lush vegetation. But often snow melts so rapidly in the first hot spring days that water rushes off the face of the frozen ground, straight into the creeks without making the journey through the soil first. Fourteen of these spring-fed, runoff-heightened streams carry the water off the mountain, feeding into the Blitzen River, where that water becomes the life blood for the dry lands below.

Not until I stood at the very top of Steens Mountain did I gain a visceral sense of what this water means for the desert. Near the rim of Kiger Gorge, one of many spectacular glacier-carved canyons on the mountain, my dog Juneau and I stood one windy day in July. We hopped across tilted rocks and slogged through snowfields, leaning hard into the wind, until we came to the edge of the mountain. I peered over, looking straight down to the desert floor thousands of feet below. Falcons carved great circles of flight on the updrafts that threatened to toss me over the edge. Signs warned nervous parents to hold on tight to their children, lest they go sailing off the rim. Beneath us, to the east, the flats of Alvord Desert spread as level and barren as a racetrack. Where Pike Creek tumbled out of its gorge and rushed onto the sagebrush below, a thin line of green zigzagged across the desert. Little else seemed to be growing beneath us, except where one enterprising rancher had tapped into the waters for an irrigation pump, and a center-pivot irrigation system watered a single fragile circle of alfalfa. Without water from Steens Mountain, farming would not survive long. But increasing demands upon that water make the future of agriculture in the region uncertain.

Where Juneau and I teetered on the rocks high above glacially carved valleys, saber toothed tigers once growled and grumbled, and dense conifer forests once spread. Now only two tiny groves of conifers persist on the mountain, remnants of the ice age, isolated by wind and drought and cold. Tiny tundra plants typical not of the desert, but of cold landscapes thousands of miles north, crouch in the rocky clefts. We balanced on what ecologists call a "sky island"—a peak that stuck up above the sheets of ice during the Pleistocene glaciation, giving refuge to species that had existed before the ice. These bits of remnant flora speak of a past when the islands were connected and animals moved freely among them. And they speak of the future as well, for current conservation efforts in the region are focused on protecting these remnant plant communities, largely because of their links to the distant past. Past and present are tangled together, high on these peaks.

Chased from the rim by rising winds and gathering storm clouds, Juneau and I ducked into a little basin above Kiger Gorge and rested by a miniature tarn—a tiny pool of water fed by a melting snowfield. Our basin held a sedge meadow, cut by a meandering stream. The water slipped by, moving across pebbles that reflected gold light underwater, the water clear and cold and slow as it traced its journey across the meadow. Down lower on the mountain, people have muddled and muddied these waters, channelized them, ditched them, drained them, and filled them with carp that wallow in the shallows like hogs. But high up on the shoulder of Steens Mountain, in clear basins, you can still see light reaching through water to the golden gravel beneath. In its journey through the meadow, the creek moved through territory it had visited in another season long before that summer afternoon. Water tells a story of its own past, and the high meadows tell another story of a more recent past, for they still bear scars of grazing wars in the early twentieth century.

As storm clouds gathered, Juneau and I followed the trickle of stream down the gradually sloping west side of Steens Mountain, hopping from rock to rock until we reached lush groves of aspen where I could pitch my tent. Generations earlier, Basque and Irish herders had followed the same route, trailing the millions of sheep that summered in the high basins, helping to create scars on the high elevation soils that are still evident. During the height of the sheep-herding boom on Steens, in the aspen groves where I waited out the summer storm, other women had waited before me. They set up big white canvas tents for the summer in meadows eventually named Whorehouse Meadows in their honor. As the storm built up, I walked from tree to tree in the grove, running my fingers over carvings left in the aspen by impatient herders, back in the days when the mountain thronged with sheep and with men and women escaping from the summer heat on the desert plains.

When the trees began to thrash about and branches to crash down overhead, I retreated to my tent, listening to the wind whip the nylon overhead and the lightning crack and the thunder smash, one right after the other and right overhead. When the storm finally passed, I sat on the edge of the swollen creek, kicking my feet in the water, watching the water churn and dip and eddy about. While the stream moved across the meadow, it was eroding its own channel, eating its own house and home away, sucking up sediments, transporting them downstream, and depositing them once more. Erosion was a constant force, pulling channels across the floodplain. But while erosion tried to push the stream wider and wider, willow roots helped to hold the banks together. The roots of sedges and willows reached down into the soil, binding some chunks of the banks together, letting other chunks get tugged off into the water. High on the mountain, where the stream was relatively healthy, riparian vegetation buffered the water's erosive force. Yet no matter how healthy the plants were along the creek, they could never prevent erosion the way a channel lined with cement could—a fact that complicates the task of people who want to save creeks by preventing erosion and other forms of change. Change is a natural part of ecosystems, yet too much change from introduced sources such as cattle can destroy much of a stream's value for other creatures.

The storm had stirred up chunks of logs, boles from the aspen, piles of twigs, and rootballs. These looked like a mess to my tidy Scots-Irish soul, but the woody debris gave the redband trout cover from their predators and also helped create little pools in the stream channel. Debris jams and dams gave me a place to hop over the creek, and they provided decaying grasses, twigs, cones, and leaves that would eventually feed the aquatic insects that sustained the trout. Stems, logs, roots, and other riparian vegetation slowed the water's journey through the basin, trapping sediment kicked loose by the cows and sheep that still grazed the high slopes of the mountain. The messiness that lay all around me acted as a sponge, slowing the floodwaters, soaking up waters that ran over the land during spring snowmelt, trapping sediments, and storing water that would slowly seep back into the stream later that summer. For generations, people had tried to clear such debris out of creeks, believing that a creek's function was to transport water and that a clean channel could do that most efficiently. Yet their efforts, however well intentioned, did more harm than good, for the more people tried to separate water from land, the more the waters dwindled away.

The willows that snagged my T-shirt, the currant bushes that tangled my hair when I ducked beneath them to make my way down the creek, the sedges that lined the marshy spots—these are all thirsty plants. They suck up water from the creek and pump it out into the air through evapotranspiration. This explains why people often mowed riparian plants down—they thought the plants were stealing water from livestock and more useful vegetation. But using water does not always mean reducing the supply for everyone else. Even while they steal water, such plants can indirectly increase the supply available to other plants, for they make the boundaries between water and land more complex. This slows water flow, keeping dirt from flooding the streams and clogging the gills of fish. Their leaves shade the streams, reducing water temperature. Their branches and dead wood fall into the water, trapping debris, forming dams, and creating deep pools of scoured gravel where fish can spawn.[1] Riparian plants help to blur the boundaries between water and land.

As I followed the Blitzen River down through a segment of public land officially designated as a Wild and Scenic River Area, I stumbled on five dead coyotes tossed across the trail, their necks sliced open, blood clotted on their fur, their paws hacked off, their entrails draining into the river. Ranchers on the edge of failure still feel threatened by predators snatching away their lambs and calves, and some lash out against that threat. But these coyotes signaled more than economic anxiety—they were emblematic of past hatreds that still have a powerful force in the basin. Anger at predators, environmentalists, and federal managers who threaten the mythic past of cowboys on the range is as strong as anywhere in the West. No matter how remote or pristine Steens Mountain seems, past and present and human and nonhuman weave together there, often in ways that are still lethal for nonhuman nature.

I followed the river to the point where it enters a nearly level valley, slows dramatically, and begins to meander across a wide floodplain in its journey to Malheur Lake. By the time it reaches the lake, the river has slowed enough to create a set of fertile wetlands. Surrounded by desert, the riparian landscapes teem with life; millions of redheads and canvasbacks and pelicans and avocets and sandpipers and snow geese and trumpeter swans rise up in great flocks that blacken the skies.

The first time I saw these marshes, I have to admit that I was disappointed. I had expected a land of tremendous scenic beauty, for my birder friends kept telling me this was their favorite place in the Northwest. But my eyes, used to the montane spectacles of the Cascades' volcanic peaks, didn't see much beauty in the harsh landscape. The light was flat and hard on a March afternoon when I first crossed over Wright's Point with my friend Rita and drove down across the rimrock toward Malheur National Wildlife Refuge. Cows grazed as far as I could see, and just a few shorebirds lifted into flight. We stopped at Refuge headquarters, ducking under the cottonwoods surrounding the old stone buildings, and gathered up maps that mostly told us where we weren't supposed to go.

The rest of the afternoon we walked along the refuge's gravel roads, peering through willows that rimmed muddy canals, flushing up ducks from the brood ponds. I had expected spectacles of lavish beauty and easy birding, and this didn't seem like much at all—a few ducks, a few little warblers in the willows that flitted off too quickly for me to identify with my pathetic birding skills, and a startling abundance of roadkill.

As the afternoon light lengthened into dusk, Rita and I headed off the refuge to find a spot where we could pitch our tents on Bureau of Land Management (BLM) land. Soon we found a little track that sliced up the side of the rimrock mesa overlooking the marshes, and we bumped up along the gash through the sagebrush and cheatgrass. Away from the lakes, when we crossed the rimrock cliffs and bitterbrush flats, I felt like I was making my way across the moon. We trudged through shadscale wastes, crossing miles of sand drifting up against salt-brush and greasewood, stumpy shrubs burnt to a crisp in the sun. In the flat light, it was easy to think that this might be the ugliest place on earth. But then we reached the edge of the rimrock cliffs and looked down to see marshes linked like jewels across the breast of the Great Basin. Just then the sun dropped down over the long marshy lakes, and in that lovely light, the waters lay before us, spread out on the basin's great wide table. The dusk filled with the conkarees of black-birds, with the cries of gulls and terns as they lifted and fell back over the low water, with the rush of waterfowl as they spooked one another into flight in the darkening night. The marshes stank of fermenting brine; they bit and prickled with hordes of mosquitoes; they shrieked with the harrier's cry; and they came awake at dusk with the calls of millions of waterbirds.

Rita and I were looking out over one of the most remote places in the Great Basin, a full day's drive from the nearest big city, a place few people have ever heard of, much less visited. Yet the birds that dipped and dove around us connected the landscape to marshes up and down three continents, from Argentina to Siberia. Malheur is one of the critical feeding sites along the Pacific Flyway, the migratory route of millions of waterfowl and shorebirds. With their flights and their songs, these birds tie continents together. When one marsh along that flyway is poisoned by agricultural chemicals or drained for new suburbs, places thousands of miles away will feel the effects when habitat loss means migratory birds can no longer make their journeys. Water and the flights of birds connect these places.

In the long low light of a cold dusk in the early spring, Malheur became a miraculous place for us: a place where we felt the earth spinning on its axis, where sandhill cranes warbled, geese honked, ibis lifted into flight, and profligate wild-ness flourished around us. At Malheur, surrounded by barren reaches of grease-wood and sage, lies a bewildering abundance of water, life, and song that has moved human hearts for thousands of years. It is a place of extremes, of stunning beauty or stunning ugliness, depending on the angle of the light, and a place of abundant resources or extreme poverty, depending on the amount of water one controls and the stories one tells oneself. Many of those stories began with the arrival of the cattle barons who ruled southeastern Oregon for decades.

In June 1872 a short, wiry young man named Peter French rode north out of the Sacramento Valley, forging the route I traced 125 years later. French was searching not for the beauty and the stories I sought but for something more concrete: grass and water for cattle. Just two years earlier, he had been only a hired hand breaking horses for the Sacramento "wheat king" Hugh Glenn. Soon Glenn trusted French enough to make him manager for his expansion into Oregon, and so, with six Mexican vaqueros and 1,200 head of cattle, French spent weeks riding across the northern California deserts, through dusty, dry lands where his cattle had a hard time finding sustenance. He continued north until he crossed a dusty ridge just west of the Blitzen Valley. What he saw over that ridge delighted him—an abundance of water in the desert. When the party stopped for the night, a discouraged prospector named Porter noticed the cook fire and came for a visit. Porter sold his few cows to French, along with the P brand. Since Porter's cattle were the only ones in the valley, French acquired informal rights to graze the land as well, to the exclusion of other cattle operations.[2] By the time French died twenty-five years later, murdered by a homesteader over contested riparian lands, he had built up those meager holdings into an empire of 45,000 head of cattle and 132,000 acres.

In the Blitzen watershed, Peter French, the six cowboys, and the 1,200 head of cattle found themselves within a watery world: a maze of streams, channels, wetlands, bogs, alkaline lakes, and lush riparian meadows—all fed by waters from the Blitzen River. Without that water, it would have been a barren desert, unable to support more than a few animals and even fewer people. With the water, it became the center of what was briefly the largest cattle empire in America, and of some of the most bitter battles among cattle barons, Indians, homesteaders, irrigators, ranchers, and environmentalists—all focused on who would win control of the riparian areas.

French never wrote home describing what he found that night, but three years later another rancher described the adjacent Catlow Valley as "one of the most beautiful valleys in southeastern Oregon, the bunch grass waving over its broad stretches like a grain field. . . . In addition to the bunch grass the white sage stood two feet high, rendering it a veritable stockman's paradise."[3] The basis of this paradise was the natural wealth offered by the wetlands and riparian areas of the northern Great Basin, a geography of basin and range where the rivers flowed not into the sea but into briny lakes. Between each fault-block range lay a basin with a moist valley where streams wandered into a maze of wetlands, creating riparian corridors and ephemeral pools that fed into great blue-green salty seas teeming with life. When droughts came, evaporation dried up the lakes into playas, concentrating the salts into a white alkaline crust. These were places of extremes and sudden contrasts—desert interrupted by snowy mountains and great, shallow, salty seas.

For thousands of years before Peter French looked down at this abundance of life, the river had moved across the entire flood plain, using a set of sinuous channels that changed from decade to decade. These riparian communities were anything but stable; floods, changes in rainfall, and changes in animal activities led to dramatic annual changes in the bottomlands. Some years the marshes were lush and green and stretched from one end of the valley to another, and the basin filled with water. Other years little snow fell on Steens Mountain, and by early summer the lowland streams ran down to a trickle, the riparian meadows turned brown, and the marshes slowly dried. Some years the water was so high that numerous pools and ponds formed in the valley, perfect for brooding waterfowl. Other years few pools formed, and waterfowl-rearing habitat was minimal. Yet because the Malheur Lake Basin was embedded in a much larger network of wetlands stretching from California to Canada along the Pacific Flyway, when droughts struck Malheur, shrinking the ponds and pools, migratory birds could find other places to rest and feed. Many of these other places are now gone.

Change was at the heart of the riparian landscapes that Peter French so admired. During spring snowmelt, the waters rise over their banks and spread over the bottomlands, irrigating lush riparian meadows where wild rye once waved six to eight feet high, and where camas and a plant the Paiute named *wada* once thrived, forming the primary plant food sources for the Wada-Eaters, the Paiute. Such floods could dramatically reshape the riparian lands, as winter storms pulled up entire forests and spring snowmelt swelled the streams, undercutting banks and reshaping sandbars. Although destructive in the short term, these floodwaters helped to create the fertility that was soon to nourish Peter French's cattle, for they saturated the soil for weeks and washed organic sediments and nutrients from the uplands over the lowland meadows.

Beaver were equally important in shaping the dynamic landscapes, yet by French's time most of the beaver had been removed from the system. Just four decades earlier, in 1826, the fur trapper Peter Skene Ogden had arrived with his party of trappers, the first Euro-Americans to enter the region. When he arrived in the basin, between sixty million and four hundred million beaver still ranged over American riparian habitats, a good many of them along the Blitzen. Beaver in the Blitzen watershed cut down trees and shrubs and built dams—as many as fifteen to twenty-five dams per mile of creek in prime habitat. Dams slowed the water flow, retained tremendous amounts of sediment and organic matter in the stream channel, and regulated nutrient flows. The great heaps of sediment provided a reservoir of carbon—twenty times the carbon in free-flowing stream sections—and this buffered nutrient flows, because the sediment piles released carbon more slowly than did surrounding areas.[4] The wetlands that beaver helped to create buffered floods and helped prolong the late-summer flow of streams. Beaver ponds in various stages of creation and decay formed a shifting mosaic of diverse vegetation communities across the watershed, playing a critical role in shaping the landscape.

In an attempt to discourage Americans from laying claim to the region, the Hudson's Bay Company's policy had been to trap the streams south of the Columbia dry, creating a fur desert. In July 1827 George Simpson stated this policy clearly, writing that the best protection from Americans was to keep the country "closely hunted."[5] Peter Skene Ogden, the trapper who opened up much of the Great Basin to white exploration, obeyed this policy to trap the region dry. By the summer of 1828, Ogden wrote of the region, "Almost every part of the country is now more or less in a ruined state," trapped dry of beaver.[6] The next spring he wrote,

> It is scarcely credible what a destruction of beaver [occurs] by trapping at this season, within the last five days upwards of fifty females have been taken and on average each with four young ready to litter. Did we not hold this country by so slight a tenure it would be most to our interest to trap only in the fall, and by this mode it would take many years to ruin it.[7]

Ironically, it was not the presence of beaver but rather the *removal* of beaver by Odgen's fur trappers that helped create the fertile riparian meadows that became the basis of French's empire. When beaver were trapped out of the system, their dams fell into disrepair, wetlands drained, and ponds filled in with sediments, creating fertile meadows.[8]

Although they liked the riches offered by beaver, most whites before French hated the Great Basin's mixture of desert aridity and watery fecundity. In 1826 Ogden spat out in his journal, "A more Gloomy Barren looking Country I never yet seen."[9] Few whites who followed him disagreed with this assessment. Although the U.S. Army had established a presence in the 1850s, no whites had tried to settle in the valley, put off by stories of marauding Paiute Indians, the area's remoteness from other settlements and transportation networks, and the nature of the land itself. Even ranchers had avoided the Blitzen Valley, for it seemed too swampy to be safe for unattended cattle.[10] Peter French, however, recognized that this water would be the center of his empire, if only he could manipulate it to his benefit. His efforts led to great success for cattle corporations, even as they eventually led to his own downfall.

PETER FRENCH AND THE P RANCH

For all its remoteness, Malheur Lake Basin was profoundly affected by a conflict between ranchers and wheat farmers nearly a thousand miles away in California. In 1871 the California legislature responded to the growing power of wheat farmers in the Central Valley by passing "herd laws," which ended the profitability of free-range cattle grazing in California. These laws required that ranchers fence their cattle in, instead of requiring that grain growers fence cattle out. Meanwhile, overgrazing and drought combined to reduce the forage available in the Sacramento and San Joaquin Valleys. As the range became depleted, and as laws restricted their profits, the California cattle kings began looking north to the Great Basin.[11]

By 1872 the Sacramento wheat king Hugh Glenn decided to shift his cattle operations into Oregon, where the ranges sounded promising and the Indians reportedly were being subdued by the army. In the Blitzen Valley, Peter French found a landscape that could, at least temporarily, support the weight of Glenn's dreams. As a 1913 company prospectus wrote about French, "Fully realizing the natural advantages of irrigation range, soil, and climate, he inaugurated his campaign of land conquest and as a result of his tireless energy, organized what is today the greatest property of the West, the famous 'P' ranch."[12] By the time the homesteader Ed Oliver murdered French in 1897, French controlled what may have been the largest cattle empire in the West.

French was not the only Californian to come into southeastern Oregon looking for open range. Between 1869 and 1873, other cattlemen, including John Devine and Henry Miller, came to claim the streams that drained Steens Mountain. These men were, in William Robbins's terms, "an imperial, flamboyant group representing large corporate interests."[13] But it was the particularities of landscape, not just powerful corporate ties, that allowed these ranchers to gain control of natural empires. Within the region, steep, abrupt fault-block topography made it possible for ranchers to set up what were essentially kingdoms in miniature. One company could control an entire basin, since the topography made it difficult for anyone else to challenge its control by trailing cattle or sheep in.[14] As the historian Peter Simpson has argued, "Each of those enterprises developed into huge, largely self-contained empires, with each of the operators attempting to gain exclusive control of the few streams that drained into the basin."[15]

Peter French, unlike many of the other cattle barons, was a progressive rancher for the times, determined not to follow the usual lackadaisical practices that some local cattlemen in the 1990s call "Christopher Columbus ranching"—letting your cattle loose in the spring and then discovering them in the fall.[16] Instead, French had his vaqueros ride close herd on the cattle, which made it much safer to run them in the wet bottomlands where untended animals might founder in the mire. French's cattle moved up the mountain in the spring, following snowmelt from the winter bottomlands up to the high peaks, eating the new protein-rich shoots of grass as soon as they appeared. Cattle thrived, but the new vegetation suffered under this regime.

French took the time to explore the watershed, scouting out springs and streams and good box canyons and better meadowlands, the better to control the region. He began fencing the best areas around springs and streams with willows and junipers, and he closed gaps in the rimrock with the abundant stones.[17] Fencing the public domain was illegal, but the southeastern Oregon cattle barons did it anyway, for better control of cattle, land, and water. The General Land Office Report of 1886–87 stated that Peter French had illegally enclosed 30,000 acres of public domain (nearly half the 67,900 acres fenced by all ranchers in southeastern Oregon).[18]

Fencing allowed French to experiment with unique stock-breeding methods. He innovated by blending two distinct ranching systems: the open-range system of Texas, with its Spanish-style cattle-driving, branding, and herding methods and Mexican vaqueros, and the much more sedentary Euro-English stock-farm techniques featuring breeding, feeding, and enclosure practices deriving from northern Europe.[19] The open-range system depended on longhorn cattle—animals tough enough to search for scarce water, survive fierce winter storms, and make the long trek to market. Such qualities, however, made the flesh tough too. Simultaneously, another method of cattle raising—stock farming—had spread west from the Midwest, thriving in Mormon Utah, northern California, and the Willamette Valley. Stock farming emphasized enclosures, winter feeding from cultivated hay, and controlled breeding.[20] Peter French adapted both methods to suit his particular landscape, using fencing, selective breeding, and control of winter hay to prosper.

FLOOD IRRIGATION

Manipulating water was the key to French's success. To control water, he needed a physical technology capable of directing its flow, and even more important, he needed the legal power to keep water where he wanted it—in his own hands, not in those of homesteaders and other ranchers. He started with the physical landscape itself, unconcerned about the law for the first decade he was in the region.

Much of the information we have about French's early manipulations of the landscape comes from court battles over water rights. As part of attempts to adjudicate water rights in the early twentieth century, the Oregon courts interviewed early settlers and people who had worked for French during his first years in the basin. These interviews are invaluable documents, even though they do rely on the vagaries of memory.

Since the sagebrush uplands did not seem particularly productive to French, he drained tule marshes, using the water to flood out upland sagebrush, which dies when its roots are submerged. With his water systems, French set out to make the uplands wetter and the swamp lands drier, and both of them better for cattle production. Prim Ortega, a Mexican vaquero who had come up from Sacramento with French on his first voyage into Oregon, testified that French had begun building dams and ditches "all over the place" long before any other settlers came to the Blitzen. Ortega added that, since people began taking water out of the river, there "ain't as much water coming through now."[21] Soon after French arrived in Oregon, according to company documents, he immediately "laid out a plan for the drainage of the swamp by a main canal and the irrigation of all lands between the foothills and the canal, putting the water on the land along the highest lines and using the canal itself as a final drain ditch for the irrigation system."[22] Ortega testified that French had "ten or twelve men working on the ditch" at one site in the valley, "building ditches and dams all over there."[23] As one employee testified in court, Peter French's "intention was to eventually drain the swamps and irrigate the sage brush."[24]

Yet French succeeded not by engineering natural riparian systems out of existence, but rather by recognizing the abundance offered by such areas in all their messiness and uncertainty. What made the Blitzen Valley so fertile for cattle raising was not just the quality of its soil or the abundance of its water. The complicated connections between these two created the abundance, and annual flooding mediated those connections. Other ranchers looked at the annual flooding of the riparian landscapes along the Blitzen and saw something messy, troublesome, and inconvenient, a natural chaos that needed to be either avoided or engineered away. French looked at these same floods and recognized that they could become the source of his prosperity.

With simple methods of flood irrigation developed by observing and mimicking the natural overflow of water onto riparian meadows during spring floods, French spread water over his drier riparian meadows in the spring, increasing the growth of native hay that he could cut and store for the winter. Streams were encouraged to overflow their banks, just as they had done in creating natural riparian areas. Rather than trying to reshape the riparian meadows into dry lands and reservoirs, French encouraged water to remain a little bit longer, to spread a little bit wider. These simple techniques of flood irrigation modified, but did not break, the connections between land and water.

Flood irrigation of riparian meadows gave French something critical in his quest to control access to grazing: a way to grow and harvest cheap hay for winter feed. French was at first the only rancher in the region to provide winter hay. Others scorned the very idea of winter feeding, believing that it made the cattle lazy and lifeless, as did, as one cattleman put it, "charity to a street beggar."[25] But even as early as 1877, according to Ortega's testimony, French began putting up hay.[26] Winter feeding allowed him to prosper during the hard winters, especially that of 1879–80, when most of his neighbors went bust. He bought them out, slowly consolidating his holdings in the valley, until he controlled nearly all the water sources and riparian areas in the Blitzen Valley.

French's methods of flood irrigation were soon copied by Henry Miller, the other successful cattle baron in the Malheur Lake Basin. Miller was born in 1827 in Germany, where he began his career as a butcher. After emigrating to the United States, he expanded his butcher shop into a vertically integrated empire of meat raising, meat processing, and meat marketing.[27] Miller expanded horizontally as well, buying land first in California, then Nevada, and finally the Silvies Valley of Oregon, until in 1895, his corporation Miller & Lux was reported to own 15,439,300 acres of land in the Great Basin, half of that in Oregon.[28] Henry Miller's operation eventually controlled most of the Silvies River valley draining into Malheur Lake from the north.

Miller's use of flood irrigation on the Silvies River illustrated some of the ways that ranchers manipulated riparian connections between water and land. Part of his success in flood irrigation was due to his close interest in, and closer knowledge of, the landscapes he sought to control. Running a solid cattle operation could not be done in the abstract for long, as many investors had tried to do during the booms of the early 1880s. Those investors had thought one could make one's fortune simply by investing so much in so many head of cattle, driving them to market, selling them, and reaping the profits. For a while, such a strategy worked, at least during years of good rain and strong markets. But what was happening on the land and with the animals was invisible to those remote investors. They acted as if land, water, rain, and cattle were interchangeable objects, understandable from a distance. Cattle could have been widgets; the land might as well have been a factory floor.

Miller managed in a very different way. While running his empire out of San Francisco, he still made it his business to learn about and watch over every single field, herd, and water supply. This close attention to the landscape was at the heart of his elaborate empire of flood irrigation along the Silvies River.[29] In letters to his ranch manager, Miller detailed the labor and attention involved in successful flood irrigation:

> You will get a competent man who will distribute the water so to get it over our meadows while the water is plentiful. At Soldier Meadows you should fix it so the water can be drained off on either side of the meadow by cutting small ditches and conducting the water further down and put in several dams and by holding it it can be spread on the black bog.

In another letter four days later to the same man, Miller gives more detail about his assessment of Soldier Meadows: "Also that meadow in front of Soldier Meadow is too wet and the hay is a poor quality. By a little work it could be put in condition to produce twice the amount of feed and the surplus water can be used to much better advantage below."[30]

Flood irrigation was hard, dirty work, requiring little technology but a great deal of skill and time. After digging supply ditches, flood irrigators would then cut openings in them to allow a sheet of water to flow across the field. Manure dams and plowed furrows would help direct the water.[31] The testimony of one of Miller's employees gives a sense of the work involved. Homer Mace testified that in 1888 he

> dug a ditch through a raise in the land and emptied it into a slough. . . . I dug right from the river bank into the slough. . . . Beside this I plowed furrows to spread the water. . . . I got water the first year I was there in 1887. It was not naturally overflowed. I just put in a frail proposition to start with; just put in some logs, rails and stuff and made me a dam; put in manure. I got water out and got plenty of water and I just kept making it better as I could.[32]

Miller believed that, by manipulating water, he could use natural processes—such as flood-deposition of nutrient-rich sediments—to improve natural landscapes, creating a hybrid of nature and culture. For example, in one letter he wrote to a manager that "at Mud Meadows the amount of money we have expended has so far done us no good. By proper painstaking effort a good deal of sediment could be washed on to the lowest and sourest portion of the meadow, more shallow ditches cut and water drained off at will and in that way we would make a splendid lot of feed and a good quality of hay; the way it is now it is nothing but worthless stuff."[33]

Flood irrigation was an imperfect tool. If the field was not perfectly leveled, as no field ever was, water would fill the depressions and just sit there, creating ephemeral wetlands, instead of spreading smoothly across the entire field. Even in the late summer, ranchers complained that they could not cross certain sloughs in their fields for fear of miring their horses.[34] To get all parts of the field wet, far more water than each plant needed would be let across the field, and that water might run off the land instead of sinking into the dirt. Miller complained constantly about such "waste," reminding his managers that "it is policy to absorb it [water] as near as possible and prevent it from running waste down on people's land who are not friendly [sic] toward us."[35] Most frustrating to an efficient manager, flood irrigation in the basin used natural floodwaters from snowmelt rather than stored waters from reservoirs. Such flood irrigation had little benefit for farmers trying to grow non-native crops such as alfalfa or wheat, species that needed water later in the season to thrive. Riparian grasses, or "wild hay," however, thrived under the early floodwaters. Early ranchers such as Peter French and Henry Miller learned to be content with harvesting native riparian hay, rather than introducing exotic grains and legumes.

After years of trying to make the water cooperate, French and Miller developed techniques that benefited from the wildness they had earlier tried to eliminate. For example, after unsuccessful efforts to drain the sloughs that had mired his horses, French learned how to use those sloughs for watering his meadows later in the summer, when irrigation was most critical. When the first spring floods came and flooded sloughs, workers would throw up a few pieces of wood and some manure at one end of the slough, trapping its floodwaters the same way a back channel traps river waters. Water from the slough would then percolate back into the surrounding fields during the drier months, irrigating native grasses from below.

Some ranchers learned to take advantage of the very wetland vegetation that others were trying to eliminate. French was the first to see that the wetlands and riparian meadows of the Blitzen, for all their danger to unwary cattle in the winter, might also prove a benefit by protecting cattle from winter storms. Others applied this reasoning to cattle management on the Silvies River, letting some tule marshes remain wet for winter protection. In 1900 Charles Cronin testified that tule land was valuable in its natural state, because "the cattle will winter on it without any hay at all. That is the way we kept the cattle in the Red S field. The tules also provide shelter for the cattle during the winter. . . . We have wind storms that are pretty hard on stock and this tule makes big shelter for them."[36]

The hallmark of flood irrigation was inefficiency, because irrigators rarely had precise control over the water's journey. Ironically, this inefficiency often led to the creation of new riparian habitats—habitats that are now critical remnants of ecosystems nearly extinct in the Great Basin. For example, the Riddle brothers homesteaded along a bend in the Blitzen River on Steens Mountain (upriver of what became Malheur National Wildlife Refuge), where the canyon opened up and the river slowed, meandering through the open basin and depositing sediments, creating rich alluvial bottomland soils (with the help of beaver). The bottomlands probably had once been laced with channels and covered with dense clumps of black cottonwood and willow. The brothers cleared this vegetation and did some simple grading in the meadows to create an irrigated hayfield and pasture. Although their preparation for flood irrigation removed native cottonwoods and other riparian shrubs, the irrigation created wet meadows, which now contain a lush community of native grasses and rare sedges—ironically, a native community that is now among the rarest ecosystems in the Great Basin. The Nature Conservancy has identified the Riddle Ranch meadows as the best example of wet meadows within the entire Blitzen River watershed, and among the best in the entire Great Basin—even though human irrigation ditches, not nature, were largely responsible for the extent of these meadows.[37]

In *Irrigated Eden,* Mark Fiege describes how irrigation projects in Idaho created a similarly complex hybrid landscape, partly human and partly natural. Idaho irrigators tried to transform nature for increased agricultural production, yet they imagined their work not in opposition to nature but as part of it. Even as they tried to make nature fit their tidy boundaries, nature constantly altered the hybrid landscapes that resulted from irrigation. Human actions in Idaho, as in southeastern Oregon, did not replace ecological systems; rather, they "shaped a strange and often baffling ecosystem."[38]

Flood irrigation in Oregon was an alteration, but one that maintained the connectivity of the riparian area, mimicking and even extending natural riparian functions. Homesteaders and other critics of ranching attacked flood irrigation on these very grounds, arguing that such waste prevented efficient manipulation of the landscape. But the real abuses by ranchers were not ecological (wasting water on riparian vegetation) but rather social (in preventing homesteaders from getting access to water). By the 1890s a few large cattle companies dominated the irrigated acreage in Harney County. Between 1889 and 1899 the county's irrigated area increased from 26,289 to 111,090 acres (a 322.6 percent increase), while the total number of irrigators dropped from 240 to 228, as large cattle companies bought out the water rights and irrigation systems of small operators.[39] While ranchers gained control of the riparian meadows and began turning the wetlands into what one observer called "giant hay ranches," settlers began attacking the very idea of flood irrigation.

HOMESTEADERS

At the same time the ranching industry was learning to manipulate the riparian landscapes of southeastern Oregon, homesteaders were trying to lay claim to these landscapes. In many parts of the West homesteaders encroached onto range already controlled by cattlemen, but in southeastern Oregon they arrived along with cattlemen, so neither group had clear prior rights to the landscape.[40] The historian William Robbins describes these homesteaders as "a trickling of subsistence settlers from the Willamette Valley who occupied small pieces of land along the streams in the northern part of the basin, especially in the vicinity of the Silvies River."[41] They—and their cattle—were largely Midwesterners rather than Californians (although many came via the Willamette Valley in western Oregon). Although they hoped to eventually make a living as farmers, most homesteaders turned to small-scale cattle raising, hay growing, and market gardening for income. This mixture of cultural traditions—Midwestern and Californian—created a diversity of attitudes toward managing cattle and land, and a diversity of landscapes as well.

In the 1870s a homestead family with little capital might hire the men and sons out for wages and a share of the hay to a large cattle operation such as Peter French's, while the daughters hired out as household labor in town. Family operations found most of their labor within the household, because they could rarely afford to hire many workers. Therefore, driving cattle to market took the husband or older son away from the operation for long periods. The wife and daughters performed much of the work on the small homestead: herding, feeding, branding, rounding up cattle, and cultivating hay and garden crops.[42]

As long as times were fairly good, the cattle barons and the homesteaders worked in comparative harmony, and early ranchers did not feel the need to organize themselves into cattlemen's associations to protect their rights against farmers. Instead, a kind of economic symbiosis developed, with each group benefiting from the other. Homesteaders were critical sources of labor for the ranchers during peak seasons, and ranch wages were important sources of capital for homesteaders trying to establish their own herds and small farms. Even though homesteaders resented ranchers' control of water, when ranchers started building dams and irrigating the Silvies River drainage in 1878, homesteaders' labor constructed the works and gates. In return, cattle barons overlooked the occasional illegally butchered steer, if settlers were careful to choose an animal in poor condition.[43]

Ranchers and homesteaders found common cause not just in their economic shared interests but, even more powerfully, in their opposition to the Paiute, who had made the basin their home for some thirteen thousand years. For all their differences, hating the Paiute gave ranchers and homesteaders a powerful sense of shared identity as whites. To lay claim to the basin for themselves, they believed they had to deny the Paiute's claim to a home and physically expel them. Ironically, however, the homesteaders found, after they had removed the Paiute from the basin, that they had thereby undermined their own economic livelihood and given more power to the ranchers. To understand the growing tensions between ranchers and homesteaders, we must examine the conflict that defined them both: the attempt to drive the Paiute from the basin, so that Euro-Americans could claim it for their own.

THE PAIUTE REBELLION OF 1878

When Peter French first arrived in the Blitzen Valley in 1872, the place had seemed nearly empty to him, and indeed it was this apparent emptiness that made the watershed such a fine place to establish an empire. French was delighted to find no fences, no prior land claims, no houses, no crops, no Indian villages—nothing to get in the way of acquiring water rights. That emptiness was not natural, however, as French had assumed, but rather it was the result of three decades of warfare against the Paiute, which had driven the Indians from the basin, making invisible their claims to the water and the meadows.

French believed the conflict between Paiute and whites to be finished, but six years after he arrived, he nearly lost his life in the last of the Paiute uprisings. When the Paiute attacked French's Diamond Ranch headquarters in 1878, the Chinese cook was killed. French managed to hold the Indians off long enough for his other workers to escape, winning him a reputation for heroism that helped him during later conflicts with homesteaders. More important, the conflict helped him gain access to prime grazing lands on what had been the Paiute's reservation. French's successes, like those of other ranchers in the basin, rested on the erasure of Paiute claims to the land.

For at least thirteen thousand years before Peter French's arrival, Native Americans had lived in the Malheur Lake Basin, yet this history meant little to the ranchers. Whites looked at the Paiute and believed they saw a people who had no fixed habitation, no material culture, no cultivation, no livestock, no homes, and no real claim to humanness. Just as Peter French and Henry Miller were learning to adapt to riparian Great Basin landscapes, over thousands of years the Paiute had done the same, developing strategies that allowed them to survive in a land of desert waters. But ironically, it was the Paiute's very adaptations to riparian life that marked them as barely human in whites' eyes. What made the Paiute so successful as a culture was also what gave whites their justification for taking over Paiute claims to the region.

Everyone who lived here, white and Indian, had to figure out how to adapt to two critical ecological constraints: variability and aridity. Some years the waters were high; other years the waters were low. Some years resources were abundant and other years extremely scarce. Peter French dealt with resource limitations by importing cattle that could eat riparian plants, turning the energy of the sun into flesh for human consumption. The Paiute, instead of relying upon a single exotic herbivore, survived by using a huge variety of riparian and wetland species—at least twenty-eight plant species, eight fish species, numerous rodents and reptiles, and many insects.[44] Yet this willingness to eat virtually anything meant to whites that the Paiute seemed subhuman. For example, one Indian agent wrote of the Paiute in 1857, "They are considered an indolent, thieving people, and those known as Diggers are of the lowest degree of the Indian race, living upon all species of insects and sometimes eating one another."[45]

Ecological variability posed particular challenges for building permanent settlements. Floods meant that near the marshes, rivers, and streams, dry land could not be counted upon to remain dry for very long, so building houses was difficult. But aridity meant that moving away from the unstable, swampy, marshy, buggy wetlands was not a particularly attractive option either, unless one wanted to haul water a long way. Peter French adapted by building levees to keep the river water from his ranch house and by mopping out the barns each spring. The Paiute simply moved their light shelters away when floods threatened. Rather than building with heavy rocks and juniper logs, as French did, the Paiute used extremely light material that the women gathered from riparian areas. They built their winter mat houses with frames of willow or aspen, which were covered with woven tules that women sewed together with wild rye plucked from the riparian meadows, threaded in a needle made from willow.[46] The doors were constructed from tules, and the floors were carpeted with wild rye. Such shelters could easily be moved closer to water during dry periods and farther from the creeks during high water.

Just as movement was French's critical adaptation to herding cattle productively in a dynamic landscape, it was the Paiute's critical adaptation to both variability and aridity. Most of the year the Paiute lived in very small family groups near creeks, springs, riparian areas, and marshes, making it possible to extract enough resources for survival without depleting the area. Family bands came together during seasons of resource abundance for communal hunts, for celebrations, and to harvest seasonal concentrations of riparian roots and seeds. These movements required that possessions, like houses, be lightweight and easy to haul away—for example, men and women wove plants into baskets, decoys, sandals, and conical hats for shade from the brutal summer sun.[47] Whites often did not recognize the utility of such objects, declaring that the Paiute were hardly human because they did not collect substantial, permanent material possessions. For example, in 1867, Major General Halleck, an army officer charged with controlling the Paiute, described them as the "most miserable and degraded savages," for

> they cultivate no land and build no houses. . . . Their only shelter, even in the rainy season and winter, is a miserable hut, covered with willow twigs, sage brush, or straw. They have only a few skins or the cast-off garments of the whites, for clothing. . . . They have very few domestic animals, and these are chiefly limited to the horses and cattle which they have stolen from travelers and settlers.[48]

Halleck's scorn for the Paiute reveals one of the most fundamental clashes in cultural visions: that between settled agriculturalists and nomadic hunter-gatherers who "cultivate no land and build no houses." Paiute claims to place were not based on fences, cultivated fields, houses, or permanent legal title, but rather on stories, memories, spiritual ceremonies, and fluid agreements among family bands.

Although whites believed that the Paiute merely lived upon the land as did animals and plants, the Paiute did manipulate the riparian areas to increase desired resources. For example, they frequently burned riparian zones because women preferred to weave baskets with the supple young stems that sprouted after a fire. They burned wet meadows in fall to increase production of root plants, to lure in animals that were attracted to the protein-rich shoots that grew after fire, and to protect their shelters from wild grassland fires.[49] Intensive digging, particularly for roots, also altered riparian areas for increased productivity.[50] For example, John Frémont, an explorer with the U.S. Army Corps of Topographical Engineers, wrote in the 1840s of the nearby Chewaucan marsh, "Large patches of ground had been turned up by the squaws in digging for roots, as if a farmer had been preparing the land for grain."[51] Although the whites scorned the Paiute for not cultivating exotic crops, such digging was a form of cultivation that increased supplies of native plants—exactly what French and Miller later attempted to do by manipulating natural flooding.

Frémont wrote of the Paiute in ways that both reflected white racism and helped perpetuate it. He portrayed himself as a scientist conducting a careful survey of western territory, but above all he was an expansionist tirelessly promoting both himself and American expansion into the West. To this end, he was intent on naming, quantifying, and categorizing what he saw; he also was careful to make notes for a profitable guidebook for future settlers. On his journeys, he conducted simple ecological and anthropological analyses, with the hope of putting farming and settlement on a rational basis. His reports offered a way to categorize the western geography: barren wasteland versus fertile farmland; good tribe versus bad tribe.

Frémont was much more observant of Paiute ways than were most whites who followed him. He was an explorer, in his own eyes—a scientific discoverer, not just an army man on a mission of conquest. Knowledge was his goal. But his knowledge was to be used to direct settlement, to guide the Americanization of the watery landscapes he studied. So, although Frémont's careful descriptions of the Great Basin Paiute gave anthropologists useful sources for describing Paiute life in the early periods of contact with whites, they were anything but neutral documents. His very attempt to be scientific helped justify conquest, for value judgments lay embedded in the notion that the Paiute were fit objects of scientific scrutiny, rather than the authors of their own stories.

In his careful descriptions, Frémont categorized the Paiute as less human than whites, a judgment that continued throughout the reports of many whites who followed him. For example, he wrote about the neighboring Klamath tribe in the 1840s, "Almost like plants, these people seem to have adapted themselves to the soil, and to be growing on what the immediate locality afforded," and about the Paiute, "Herding together among bushes, and crouching almost naked over a little sage fire, using their instinct only to procure food, these may be considered, among human beings, the nearest approach to mere animal creation."[52]

A decade after Frémont's accounts of the Paiute, the U.S. Army began to establish a presence in the Malheur Lake Basin. In April 1859 General William S. Harney sent Captain Henry D. Wallen to survey a shorter road from Salt Lake City to The Dalles.[53] This first army expedition marked the beginning of a federal presence in adjudicating growing tensions among Paiute bands, miners, ranchers, and settlers. Within a year the army began to focus on hunting down the Paiute, not simply categorizing them. With the beginning of the Civil War, the federal government temporarily withdrew troops from the Indian conflicts on the Oregon frontier.[54] Pro-slavery advocates in Oregon soon organized into the Oregon Volunteers, who turned to undeclared war against the Paiute. In 1864, to protect the travel routes of miners in eastern Oregon, the Volunteers began killing any Paiute "hostiles" they saw, after defining essentially all Paiute as hostile. Volunteer captain George B. Currey illustrates the ways such whites looked upon the Paiute. Currey loathed the Paiute for keeping a potential home for civilized families out of his grasp. Angered when he was ordered not to organize a winter campaign to destroy the Indians, he wrote,

The camps were abandoned, and Eastern Oregon again yielded up to the sway of the savages. And now, as in the past, two-thirds of the State over which your title extends cannot be peaceably traveled through by the citizens of the State, and a region capable of giving thrifty and pleasant homes to thousands of civilized families is reserved for the theater of the horrid orgies and murderous rites of the fiendish savage.[55]

Currey's mutterings of Paiute "orgies and murderous rites" reflected the intensifying attempts of whites to justify their own usurpation of the basin by claiming that "savages" had no right to a place capable of giving homes to civilized families.

After the Civil War, Lt. Col. George Crook took command of the district of Boise, with orders to subdue the Paiute once and for all in the region. Crook began an aggressive campaign, ordering his soldiers to proceed by "killing all the Bucks and taking none of them prisoners." By the time he had finished, he bragged that "over half the Indians were killed and the remainder reduced to a state of starvation."[56] Part of the treaty the defeated Paiute negotiated with Crook stipulated that the federal government would create a reservation for the Paiute on their ancestral lands but that the Paiute would not be required to stay there; they could hunt, fish, dig, and trap anywhere on their ancestral lands.[57] The government created the Malheur Reservation to defuse tensions between settlers and Paiute, hoping thus to free up land for ranching.[58] But the reservation quickly became a site of constraint and anger for the Paiute, as ranchers pushed onto the grazing lands and Indian agents refused to give the Paiute food, attempting to force them to stop their wandering ways and become settled farmers.

Although the first agent, Samuel Parrish, was well liked by the Paiute, in 1876 he was replaced by a political appointee, William Rinehart, whom the Indians opposed. They believed Rinehart to be fundamentally hostile to them, for he had fought against them in Crook's campaign. In his 1876 report, Rinehart noted,

> Twelve years ago it was my fortune to be in the Army and with the troops then operating against these Indians, and a more abject race of beings it was never my lot to behold. The best lodges I saw during two whole summer campaigns consisted of only a few sagebrush set up in a half-circle, as if to keep off the wind. All we found were abjectly poor, many being absolutely naked. . . . Hotly pursued all the while, they had no time to manufacture matting for the lodges, and in this condition they merited the hated appellation 'Snakes,' absolutely living in the grass.[59]

Rinehart's contempt was not limited to the Paiute. However much he hated their culture, he felt he had a job to do and that the government was refusing to let him do it. He was furious at federal officials for their stinginess and refusal to provide promised appropriations. And he was equally furious at the settlers, who thought they could murder Paiute with impunity. He railed against settlers in the basin "who deem it their high privilege to shoot at sight any Indian they may find away from his reservation."[60] When President Grant ordered the north shores of Malheur Lake opened to settlement in 1877, Rinehart understood that these marshes were a critical source of *wada* seed essential for Paiute survival and that their loss meant the threat of hunger for the tribe.[61] Rinehart saw starvation settling in among the Paiute and understood that the fault was not theirs, for it had become dangerous for any Paiute to leave the reservation to hunt or gather at their traditional riparian sites, as allowed by the treaty. Legally, they could continue to live off the reservation, but in reality they were being murdered by whites off the reservation and starved on the reservation. Although Rinehart was aware of the indignities to which the Paiute were subjected, his sympathies were limited, as he condemned them for continuing "to roam the country at will, in defiance of the wishes of the whites and in disregard of the regulations of the department providing them homes on reservations."[62]

Rinehart was frustrated not only with the Paiute but with ranchers such as Peter French who encroached on reservation land. In 1878 Rinehart complained that "stock-men are driving cattle to graze upon the lands of this reservation, and the growing dissatisfaction of the Indians resulting from this cause is likely to produce future trouble."[63] Yet his complaints against the ranchers were ignored in Washington, D. C., and nothing slowed the illegal trespass onto the reservation.

These tensions among ranchers, settlers, Indian agents, and the Paiute soon exploded into violence. In 1878 the Bannock Indians in Idaho rebelled, and nearly all of the Paiute on the Malheur Reservation joined them in the uprising.[64] Chaos ensued when army troops made the mistake of driving the Indians into the settled portion of the valley instead of away from white settlements. Much to the whites' surprise, the Paiute won the first three skirmishes against the army and white citizens, and in Rinehart's words, this "was the signal for a panic, which involved the whole settlement."[65]

The uprising was brief, yet retaliation against the Paiute was brutal and swift. That January federal troops forced the Paiute survivors to march 350 miles through the snow to the Yakama Reservation, home of tribes long hostile to them. Peter French and other ranchers quickly moved livestock onto the reservation, using the uprising as an excuse to claim the best water sources and riparian meadows for grazing.[66] With reservation lands as their base, these men became even more powerful as cattle barons, controlling hundreds of thousands of acres of prime grazing land. On March 2, 1889, the president abolished the reservation, restoring the remaining lands to the public domain.[67] As the historian William Robbins writes,

> Cattle interests and farmers subsequently forced the Indians to liquidate title to most of the valuable grazing land in the basin proper, thereby opening the former reservation lands to the full force of the market. Several decades later the Indian Claims Commission suggested that during the reservation period, settlers and stock growers had 'practically pre-empted the reservation land right from under the Indian.'[68]

Over thousands of years, the Paiute had evolved a set of cultural responses to riparian boundaries that allowed them to coexist with the change and uncertainty inherent in the desert wetlands. They took advantage of riparian boundaries, manipulated riparian fertility, and transformed riparian landscapes in ways that benefited their own culture—exactly as Peter French and Henry Miller were trying to do. When the whites "came like a roaring lion"—in the Paiute woman Sarah Winnemucca's phrase—thousands of years of cultural coherence were fragmented into new patterns: new political alliances, modes of governance, chiefs, social bands, migrations, and terrors. The Paiute went from being hunters to being hunted; from being gatherers to being gathered up onto reservations. Their claims to the landscape came crashing up against the single potent story of the whites that made the Paiute the enemy of progress.

Ranchers and homesteaders joined in destroying the Paiute, both groups telling a story that justified their actions by claiming that the Paiute had no right to the basin because they weren't cultivating the soil, using water to irrigate crops, or building permanent homes and communities. Yet within a decade homesteaders were making exactly the same accusations against ranchers. By agreeing on a story about the Paiute, ranchers and homesteaders found a shared identity, but after the Paiute uprising their cohesion disappeared along with their common enemy.

CONSOLIDATING THE LAND AND WATER

Although whites were sure that Indians were the enemy, they had failed to realize how much they had benefited from creating such an opponent. Both ranchers and homesteaders had depended on the army as the surest market for their beef and hay. When the army posts were abandoned after the 1878 rebellion, local markets for these products vanished. Ranchers suffered less, for they could afford to drive cattle to the railroad over two hundred miles away. As Margaret Lo Piccolo writes, "Clearly the situation heavily handicapped small stock raisers. The large corporations that initiated drives and marketed their own herds acted as the only buyers in the region during the decade."[69] Cattle barons bought cattle from the small owners each year, but at lower prices. During hard times, they might not buy at all. "A large owner could ruin a small one simply by refusing to buy from him," Peter Simpson argues.[70]

The tenuous peace between ranchers and homesteaders disintegrated in the hard times of 1879–81, when the closure of army bases and the temporary collapse of national beef markets combined to hit settlers and ranchers hard. Added to this came the harsh winter of 1879–80, which wiped out many of the herds, especially those of small owners who had failed to put up hay. The winter was one of the most severe on record. Heavy snows were followed by ice crusts too thick for cattle to scrape away, which meant starvation for the thousands of cattle lacking winter hay. French's cattle in the Blitzen survived fairly well, because he fed them. But throughout the rest of the basin, many cattle died and small operations collapsed, increasing tensions between cattle barons and settlers.[71]

French and other cattle barons came to believe that control of legal title might be as important as control of the physical landscape. Buying out failed settlers was the only legal way for French to acquire land, and his readiness to take advantage of others' failures led to increasing tensions. In the Blitzen River drainage, 85 percent of those who came during the 1870s were gone by 1890, most having sold out to French. In just four years, between 1887 and 1891, the French-Glenn Livestock Company gained deeds to 14,813 acres from settlers.[72] Of fifty-eight families who came into the Diamond and Happy Valleys during the 1880s, only three appeared on the rolls in 1890—a sign of the economic and social instability of the place.[73] The rancher David Shirk (a sworn enemy of Peter French, for they had competed for the same wife) commented in his autobiography, "Immigrants of that day were born in a wagon, here today, there tomorrow."[74] Sales by settlers accelerated during cycles of dry years, such as the late 1880s. In February 1890 the *East Oregon Herald* begged settlers not to be discouraged: "Harney has a great future, and men are riding over lands today, and are leaving ranches they cannot come back to."[75]

Consolidation of large holdings by cattle barons was tremendous during the late 1880s. French began gaining possession of riparian meadows and uplands by any means possible, some legal, some less so. As a 1913 company document bragged,

If the very cream of the country was not picked up it was not the fault of Peter French. He proceeded first by acquiring title under the Swamp Land Act from the State of Oregon, and in this way secured those miles and miles of swamp lands, the drainage of which has been completed by the present management, so that thousands of acres now await the plow, and most prolific crops of barley have been grown, where only a few years ago the cattle used to mire down in the tule swamp.

Following his policy of securing the water, French, as fast as lands were surveyed by the government (which surveys he took steps to expedite), secured in
one way and another the legal title to the entire Blitzen River from its source to
its mouth, and of every tributary and of every spring on Steens Mountain that
had a perpetual flow. . . . Thus Peter French, sometimes by scripping, sometimes
by purchase, sometimes by the Desert Act and the Swamp Land Act, secured every
foot of valuable land between foothill and foothill.[76]

French hired dummy entrymen to establish homestead claims along streams and
then sell the homesteads back to him, transferring the water rights along with
the acreage. He also encouraged all of his workers to take up homestead claims
so that he could then buy them out. For example, Prim Ortega filed on the original P Ranch homestead, then sold the homestead in 1884 to French.[77] Lo Piccolo's
analysis of company land records shows that between 1882 and 1889, French
gained 26,882 acres, of which 16,097 came from employees listed in the company ledger of 1883–86.[78]

French's major holdings came through manipulations of the Swamp Land
Act. Designed to fulfill a national vision of transforming wasteful wetlands into
prosperous farms, this act instead fostered an empire not of turnips but of cows.
In 1860 Congress extended the original Swamp Land Grant to Oregon (and other
states). The law granted to the State of Oregon "all swamp and overflow lands,
unfit for cultivation, within its boundaries."[79] The goal was to encourage cultivation by settlers, not monopolies by a few corporations, but monopolies soon
resulted.

The riparian geography of southeastern Oregon made this act particularly
significant for cattlemen, for the valuable lands in the region were what the act
called "overflow lands"—riparian meadows flooded by early spring runoff, which
produced abundant crops of wild hay. Ownership of this land meant control of
not only riparian fertility but of the sources of water, which allowed ranchers
to monopolize the "countless acres of semi-arid rangeland dependent on this
water supply."[80]

By 1870, with cattlemen coming in from California and the extension of a
railroad to Winnemucca, Nevada, state legislators in Oregon decided to take steps
to secure this desert swamp land. The legislature directed the state to select and
publish a list of its overflow lands. Any citizen could apply to purchase the listed
lands for $1.25 an acre. The applicant had ninety days to pay 20 percent of the
total price and had to pay the rest after proving "cultivation in grass."[81] By 1872
applicants had applied for a total of 5,828,715 acres, an extraordinary figure for
a state that eventually sold only 5 percent of that amount.[82]

Because the land had not yet been surveyed or investigated by federal officials, the General Land Office withheld certification, making it impossible for the state to act upon the claims. Nevertheless, the state accepted down payments for 215,000 acres and sent "certificates of sale" to many of the applicants, who therefore assumed that they had bought the land and proceeded to occupy it.[83] Not until 1891 did the secretary of the interior approve some of the tracts for classification as swamp land—some twenty-one years after federal law declared that such land would be given to Oregon. When the federal land patents came, they rarely were for exactly the same lands the state government had approved, requiring relocation of fences and acerbating tensions between settlers and ranchers who tried to claim the same land.[84]

When the dust settled from the confusion of the Swamp Land Act, a few large cattle companies (mostly funded and controlled by California corporations) owned much of the region. As the federal geologist Gerald A. Waring stated in 1909, "Practically all of [the marsh lands] are controlled by the larger stock owners."[85] A senate committee of the legislative assembly of 1887 investigated the records of the Land Office, finding that those of sales between 1870 and 1878 were almost nonexistent, so that an accurate tracing of swamp claims was impossible. The committee also discovered that the large cattle ranches of the southeastern part of the state had benefited the most from the swamp grant.[86] For example, in 1877 Peter French had acquired 48,570 acres in one sale, by buying out A. H. Robie's holdings of "swamp and overflow" land in the valley acquired through the Swamp Land Act. Henry Miller, it was rumored, had claimed his vast acreage of swamp land in Oregon by riding in a boat over his claimed "overflow" lands to prove they were indeed swamp lands—but the boat was carried overland by a wagon pulled by mules.[87]

Flagrant abuses of the Swamp Land Act drove settlers to action against the cattle barons. They gathered evidence of fraud and asked state officials to revoke title and open these lands to homesteaders. In his 1886–87 report, the secretary of the interior found that of forty-five swamp land selections made in Oregon in 1883, thirty-eight were dry. The secretary noted what he termed the most "unblushing frauds practiced in Oregon," which "gave control of lands around lakes and watercourses to a few individuals."[88] On March 14, 1889, the *Harney Valley Items* reported allegations that Swamp Land Act claims were "being manipulated in the interest of land syndicates and monopolies." A month later, the same paper reported, "The land office has issued special instructions to Special Agents Elliot and Armington to re-examine some 58,000 acres of land in Harney Valley, Oregon, alleged to be not swamp."[89]

There was much talk in the local newspaper about the plight of farmers: "We are getting poorer each year; our masters are becoming richer, yet they produce no wealth except the wealth of a glib tongue."[90] One settler wrote in 1887,

I want to express my gratitude to the *Harney Valley Items* for the position it has taken in its utmost endeavors to show to the people at large the great fraud that is being perpetrated upon this patient and hard-toiling community. The settlers of this valley can never repay the *Items* for the noble stand it has taken. . . . It has fought unflinchingly for their cause in the darkest days of their troubles.[91]

Although a few of the cattlemen's claim to swamp lands were declared illegal, most appeals to the state failed. Settlers then took the cases to court themselves. Litigation lasted for years; settlers won a few of the cases, but simply ran out of money contesting the others.[92] The cattle companies firmly controlled the valuable riparian lands in the desert. As William Robbins argues,

During those years, the cattle industry in southeastern Oregon operated in a Darwinian economic and ecological world, playing footloose and free with federal and state land laws, gaining a stranglehold on water rights, buying out smaller owners to expand their holdings when it was to their advantage, and keeping too many cattle on the ranges when the market was soft.[93]

OVERGRAZING

In the wake of intensifying competition, both cattle barons and homesteaders tried to increase profits by maximizing their livestock holdings. The result in southeastern Oregon, as across the West, was overgrazing, the effects of which still scar the riparian areas and uplands a century later.[94]

Overgrazing resulted from political as well as ecological conditions, and it helped spark political as well as ecological change. In 1875 the Central Pacific Railroad line completed a shipping facility at Winnemucca, Nevada, south of Malheur Lake, giving cattle barons relatively rapid access to San Francisco beef markets.[95] With an efficient transportation infrastructure in place, ranchers loaded the open ranges with more animals. Peter French increased the cattle he ran on the P Ranch, until by 1885 he had 45,066 head of cattle.[96]

Overgrazing was intensified by the collapse of beef markets. When prices were low, few young cattle were sold, and herd sizes rose while ranchers waited for better prices. The federal government responded with the Gordon Report, an important study that did a fine job documenting the extent of damage but failed to recommend any regulatory or ecological solutions.[97]

The Gordon Report was motivated in part by the disastrous winter of 1879–80, when extraordinary cold led to high cattle mortality across the West. Clarence Gordon wrote that "The ranges at the beginning of the winter of 1879–'80 were more heavily stocked than ever before. . . . Grasses had been also injured by the summer's drought." He singled out Steens Mountain in the basin as a site of particular damage, noting that that "The Stein's [sic] mountain region was overstocked at the beginning of 1879."[98] Overgrazing had depleted the forage enough, the report claimed, that cattle were suffering, failing to gain weight. "The general explanation of this failure of beef animals to lay on fat as readily as formerly is that the overstocking of ranges has so generally occurred as to cause injury to the bunch-grass, white sage, and greasewood, which constitute the winter pasturage."[99]

Overgrazing meant that riparian meadows became even more critical, and even more stressed, for "cattlemen looked for supplementary food supplies and they found them in the . . . wild hay meadows especially numerous in southeastern Oregon."[100] Attempts to "improve the range"—as ranchers fenced riparian areas in, converted them to alfalfa, and hayed them for winter food—took some of the pressure off the land, but only for a short time. Ironically, the Gordon Report noted, these efforts soon helped to degrade riparian areas. The result, as the 1883 edition of *West Shore* magazine reported, was a landscape "almost bare of grass except for a few clumps under the dense scraggly sage brush."[101]

In the wake of the 1879–80 disaster, cattle and sheep populations rebuilt, until a combination of dry summers and cold winters—"the worst nightmare of cattlemen"—occurred in the late 1880s.[102] Cattle prices collapsed in 1885 and 1886, and ranchers held their stock from market, hoping for higher profits. Drought during the summer of 1889 led to disaster. As the *East Oregon Herald* reported, "The Malheur and Owyhee Rivers sank low in their channels at the spring flood time and by August the Donner and [sic] Blitzen dried to a few puddles in its bed."[103] The hay crop from riparian meadows was extremely light. When the ranges southeast of Steens withered away from the drought that July, ranchers such as John Devine dumped three thousand head of cattle on the impoverished ranges of Harney Valley.[104]

By January 1890 the *Harney Valley Items* reported that "large numbers of cattle are suffering for want of feed, and enough are dying to make a big hole in next year's rodeos. . . . Cattle are suffering. . . . Not having hay for them the poor beasts began eating each others' tails off." Week after week, the papers recorded a worsening situation. At the end of January the *Items* glumly noted, "Money, fuel, and hay scarce. . . . Range cattle reported so poor that death will claim two thirds when the thaw comes."[105]

During the drought, a flood of sheep into the basin intensified conflicts over dwindling resources. Sheep had been grazed in Harney County nearly since the introduction of cattle, but they had not become a major economic force until 1895, when cattle prices dropped and sheep prices rose, attracting larger bands to the unappropriated rangelands high on Steens Mountain.[106] These were among the last free ranges in the Pacific Northwest, and so when Columbia Basin ranges began to close to sheep with the shift to wheat production there, many transient herders moved south to Harney County. Although some long-established cattle barons such as Henry Miller converted part of their operations to the more profitable sheep, most sheep were herded by transient, often foreign, herders who owned no property and did not pay taxes in the county.

In part because of the sudden influx of sheep, arguments over an equitable range policy raged in the county, with positions divided along class lines. Newspapers favored homesteading; large cattle operations favored leasing; and smaller operators favored federal regulations. Transient sheep herders were an easy target for all.[107]

Many cattle barons blamed sheep for damage done by their own cattle. For example, F. C. Lusk, the man who took over P Ranch operations in 1898, wrote that sheep "destroy the willows and small brush on the little mountains streams, etc. which causes the snow to go off with a rush and consequently there is not water to irrigate meadows."[108] Although Lusk was right about overgrazing damage to riparian areas and the resultant damage to the larger landscape, he was wrong to blame only sheep and not the tens of thousands of cattle that he himself controlled.

Even government experts fell into the trap of blaming sheep for all the grazing damage. Gerald Waring wrote in 1909,

> Within the last few years the high price of sheep and the low price of cattle have led to the introduction of rapidly increasing numbers of sheep. As in every other grazing region that they have entered, sheep are rendering the none too abundant range unfit for cattle and horses, except in those portions of the high desert from which the scarcity of water excludes them; and already in the mountain regions the results of overgrazing are very apparent.[109]

By 1901, when the Bureau of Plant Industry sent out an inspector, David Griffiths, to report on grazing conditions, Griffiths estimated that at least five hundred sheep per square mile were grazing on the four hundred square miles of Steens Mountain pastures, and staying there for at least four to five months of the year. In other words, 182,500 sheep were on the mountain, for a total of nearly a million sheep months that year. As Griffiths wrote in his report, "To say that the southern portion of the region is overstocked would be putting the matter very mildly.... On the whole trip of three days we found no good feed, except in very steep ravines."[110] In some areas, "there was practically no more feed than on the floor of a corral. We passed two areas at least 2 miles in extent in which even the surface of the ground was reduced to an impalpable powder."[111]

Griffiths noted that intense competition for early season grazing led to particular pressure on riparian areas. At higher elevations

> the destruction of the [riparian] shrubbery, all too scanty in this region, had a potent influence on the lowland meadows and the mountains themselves, both in relation to the conservation of moisture and the protection of the surface soil from the erosive action of water. The destruction of the vegetation means vastly more than simply depriving cattle of food."[112]

Irrigation of sagebrush uplands for alfalfa meant that little water was available for natural flood irrigation of the lowland meadows. Competition for water, and particularly irrigation of hay meadows for winter feed, had reached the point that "the rivers do not reach the sinks, and the small tributaries from the mountains often do not reach the main channel, for all of the water that succeeds in getting down to the fertile sage-brush areas near the river bottoms is used . . . in irrigation."[113]

Griffiths reported that the conflicts between sheepherders and cattlemen had become "as in the majority of the open range regions, often a very bitter one." Local cattlemen were furious at the sheepherders, for "All of the water on the fertile lowlands was taken up in early days by cattle interests, and the cattlemen looked upon the use of the mountains for grazing purposes as a natural right." Yet, as Griffiths argued, cattle rather than sheep did the most damage to riparian areas.[114]

On private lands, riparian areas were in poor condition, for they were not only overgrazed but were also mowed for hay, and, as Griffiths grumbled, "It is needless to say that these areas are taxed to their full capacity. . . . A piece of ground from which a crop of hay is removed during the summer will not usually maintain its productiveness in any region if every particle of vegetation remaining is pastured off during the fall and winter seasons." But unregulated competition meant that riparian areas on public lands were in even worse condition, Griffiths reported, for "these areas are always closely grazed and present a very unpromising appearance. No open-range lowland was seen on the whole trip which had much feed upon it excepting that consisting of the tough and persistent salt grass. Everything else had been cropped closely."[115]

Griffiths concluded by warning that overgrazing was ruining the very source of the region's prosperity. The problem, he argued, was not due to climate, but to overstocking—which in his view was fundamentally a political, not just an ecological issue. He wrote, "It does not appear clear how matters will improve in this respect in the near future as long as there is no inducement for anyone to do aught but get all he can out of the little that the country does produce."[116]

In other words, because ranchers' only reward was to get as much for themselves of what little remained, no one could afford to conserve.

The ultimate solution? To provide more grass, so there would be less competition for the resource, Griffiths reasoned. But how to get more forage? Draining the wetter basins would furnish more productive grass and fewer of the native riparian plants that cattle don't like, such as rushes, sedges, and wire grasses.[117] The irony is that these methods were both part of the problem and part of Griffith's proposed solution. But a still larger irony, which he did not seem to realize, was that these solutions would not address the fundamental political problems at the heart of the overgrazing problem: lack of regulation and inducements toward unrestricted competition. His suggested solution—to decrease competition by increasing forage—assumed that livestock numbers would remain steady as grass increased. But because many operators were competing for the same grass, livestock numbers would—without political regulation of some sort—increase along with the grass, resulting in continued overgrazing.

Even with federal documentation of overgrazing, little was done to control the problem. Lack of regulation of grasslands; confusing land tenure regulations; and conflict over swamp laws, water rights, and unfenced range—all of these magnified tensions. For the larger ranches, more profitable management meant tighter control of water and riparian areas, and that control threatened what new homesteaders needed as well. Cattle barons felt they had few legal opportunities to control the forage and water they needed for expanding operations, so they felt justified in manipulating and sometimes simply breaking the laws. Their tactics against settlers became increasingly ruthless.[118] When dry weather and depression returned in 1893, the conflicts exploded in open warfare.

2 / Conflicts between Ranchers
and Homesteaders

S outheastern Oregon became famed as the stronghold of the largest cattle
empires in America and also as seat of some of the most brutal conflicts
between cattle barons and homesteaders. The Blitzen Valley witnessed
explosive battles between these groups, battles that derived much of their feroc-
ity from differing interpretations of nature. Both groups tried to transform the
abundance of riparian areas, hoping to bring about a more productive, orderly
nature. Their attempted transformations were driven by conflicting stories that
delineated the ideal relationship between land and water. Finally, in 1897, one
of the most powerful cattle barons of the West was shot by a disgruntled homesteader in a conflict over riparian boundaries.

The wealth of cattle empires depended directly on the wealth of the riparian
meadows. One key to the success of Peter French's empire was his ability to
manipulate riparian structure and increase the action of natural floods without
entirely destroying riparian function. But his control of water was only partial,
and powerful social and ecological tensions destabilized that control.

Eventually, by the end of the era of the cattle barons, what was once the greatest cattle empire in the West became the greatest bird empire—what some called
an "empire of nature." Yet *all* of the empires in this valley—the cattle kingdoms,
irrigation empires, and empires of ducks—were at heart empires of nature, for
they all depended directly on the abundant natural resources of desert riparian
areas for their wealth.

During the 1890s a populist, antimonopolist rhetoric emerged among settlers
and news editors. This rhetoric was not unique to southeastern Oregon: droughts,
economic depressions, and political shifts had stimulated populism throughout
much of the nation. Small farmers were convinced that cattle barons, railroad
corporations, and land speculators were colluding to destroy the working man.
As the social historian Peter Simpson writes, "Political radicalism was the inheri-
tance of the farmer of the 1890s . . . a radicalism tinged with desperation."[1]

These tensions were inflamed by news editors throughout Oregon. The editor of the Portland magazine *West Shore* wrote,

> One cannot peruse the foregoing review of the location, size and condition of the numerous valleys of [Harney] county without being forcibly impressed with the fact that the stockmen have appropriated the lion's share. . . . They have thus far been successful in keeping settlers out and appropriating the whole country to their own use, even completely fencing in many of the valleys, and grazing their vast bands of cattle free of expense on the public domain. In this every taxpayer in the county has been wronged.[2]

The local radical paper within the basin, the *Harney Valley Items*, deplored the fact that the great western ranges were passing into "the hands of a few big cattle or sheep companies" and predicted that soon "an aristocracy of range lords and cattle kings would rule our mountains and plains."[3]

MEANDER LINE CONFLICTS

As early as the 1880s, the water levels of Malheur Lake had begun to fall as more water was diverted for irrigation, and the lands between the high water mark (the meander line) and the waterline became contested territory. The central conflicts rested on what the boundaries between water and land meant in a place where those boundaries were never fixed, and how those shifting boundaries affected legal title.

Above the meander line was French's land, but between it and the actual lake levels, ownership was uncertain, because the precise nature of that riparian landscape was uncertain: Was it water or was it land? Large operators such as French claimed rights to those newly formed spaces because, in their view, it was not land but lakebed, and therefore part of their original riparian claim. Homesteaders, on the other hand, argued that the new spaces were true land, and therefore should be considered public domain open for settlement.[4] The State of Oregon in turn claimed the spaces for itself, arguing that they were neither private (i.e., French's) or federal public domain, but rather state land under the terms of the Swamp Land Act of 1860.[5]

As settlement increased in the basin, human interventions made the lake bed increasingly unstable, thereby decreasing the stability of legal title as well. More water diversions from the Blitzen and Silvies Rivers were built for irrigation, which lowered water levels further in Malheur and Harney Lakes. In the spring of 1881 the sand reef separating the lakes broke (whether through natural forces or through the angry kick of a cowboy's boot remains unclear) and the waters rushing from Malheur into Harney Lake cut a new channel two feet deep. Because Harney Lake's elevation is slightly lower than that of Malheur Lake, the new channel lowered the level of Malheur irrevocably, exposing more lake bed for people to fight over. The channel not only lowered the lake level but also made it more unpredictable. In some years, erosion led the channel to silt in, blocking flow from Malheur into Harney and so raising water levels in Malheur, swamping out the squatters. Other years heavy spring runoff washed out the silt in the channel so that Malheur's level dropped precipitously, even though rainfall was high—exactly the opposite of what people expected.[6]

While the disputes over title staggered through the courts, squatters began to settle on the exposed lake bed, first grazing a few cows, then trying to grow alfalfa and grain. When homesteaders first moved onto this unsurveyed land in the early 1880s, Peter French raised no objections. Settlers who already owned land and wanted to increase their ranks in the social battles brewing against ranchers began to urge landless farmers to settle on exposed lands, even given uncertainty about title and lake levels. For example, a letter in the *East Oregon Herald* from a Mr. T. V. B. Embree of Round Island Farm called the public's attention to the body of

> country embraced between the rimrock, the lakes and the Narrows, all lying west of the Narrows. This is all unsurveyed, and there is over a township of it, much of it appears on our maps as "lake," and at times, it is overflowed, but is certainly the best body of meadow-lands in Harney valley. . . . This country will one day make the great hog growing region of Harney valley. All that is wanting to make it a thrifty land is intelligence, industry, and "grit" to go to the ground and open up.[7]

In the late 1880s settlers began building shacks on this newly exposed land and even lived in them, except when waters were too high.

French soon tired of the incursions of squatters and began to fight them in court. Between 1890 and 1895 he instituted three eviction suits, claiming the space on the basis of riparian doctrine. Squatters fought him in court by attacking riparian doctrine itself, as well as his original title, trying to show that his claims to the lake bed had been acquired by fraud through misuse of the Swamp Land Act.[8] By 1895, in response to these court cases, federal surveyors made two separate surveys of the lake, and five major cases had gone to court.

The courts had enormous difficulty deciding on legal title for two major reasons. The first was legal: Oregon had still not codified its water law, and courts were still debating between riparian doctrine and prior appropriation. The second reason was ecological: conditions in the riparian landscapes were fluid and uncertain. As Ted Steinberg argues in *Slide Mountain,* this fluidity makes it difficult for courts to fit such landscapes into any legal doctrine that requires a clear separation between land and water.[9] Oregon water law started with a foundation in English riparian law, which had grown out of a common-law tradition of riparian rights. Riparian rights gave whoever owned the land along a river's banks rights to the river's water, as long as use of the water did not "diminish its flow, alter its course, or degrade its purity." But even though French and other cattle barons claimed their water rights on the basis of riparian doctrine, that doctrine had been changing for quite some time.

In Britain and the eastern United States, when the Industrial Revolution led to the increased development of water power to drive the engines of industrial capitalism, riparian rights had proven unworkable as a water policy. Power development did alter the river's course, flood out owners above, deprive owners below, and degrade the water—changes not allowed under riparian doctrine but necessary for industrial development. In the eastern states, courts had modified riparian rights, allowing changes to the water's course as long as they constituted "productive use." A new body of appropriative rights, in other words, emerged in American water law, as Donald Pisani has explored in great detail.[10]

In the West, riparian doctrine had first been challenged by miners needing vast quantities of water for hydraulic mining and sluicing gravel. They had argued that "whoever first put water to productive use acquired a permanent right to it."[11] One lost one's water rights if one didn't continue "beneficial use." During the 1880s Colorado had developed this into formal water law by laying claim to all surface water within state boundaries and then nullifying riparian rights and enforcing water rights acquired by prior appropriation. If someone who had come first decided to use all the water in the stream, landowners downstream could not do anything about it, and later settlers were out of luck. The doctrine of "prior appropriation" became known as the Colorado Doctrine and was, in one form or another, eventually adopted by nine western states.[12]

Prior appropriation, as the historian Norris Hundley argues, "endorsed swift commandeering of water resources and rapid economic development, and it gave no advantages to communities over individuals. . . . this situation encouraged individual and corporate tendencies to monopolize as much of it as possible."[13] Yet small farmers in the Harney Basin, as throughout the West, did not see prior appropriation this way. They believed firmly that riparian doctrine favored cattle barons and large landowners, and that appropriation was the people's law.[14] Oregon water law was not able to resolve these contested points until 1909, when the state developed the Oregon Water Code in the wake of the Reclamation Act.

In 1899, the state supreme court handed down a decision in favor of the French-Glenn Company, but not on the basis of the company's claim to riparian rights. The court disallowed such rights and repudiated riparian doctrine as a basis for future claims. Instead, it awarded the company the land based on prior appropriation and beneficial use, since French had first used the water. Squatters quickly filed an appeal, and the case made its slow way to the U. S. Supreme Court.

While the legal battles dragged on, both sides lived side by side, bitterly opposed to each other. Anger fermented into a toxic brew of hatred and violence. French became a target for arsonists who tried to ruin him by burning his stacks of winter hay. The *East Oregon Herald* reported, "Peter French lost two stacks of hay by fire; supposed to be the work of an incendiary."[15] Papers as far away as the *San Francisco Examiner* mentioned the arson.[16] A letter in the *Harney Valley Items* went so far as to urge settlers to violence:

> Wealthy despots should not go unwhipped of justice. It is well known much of our land matters are in a deplorable condition; and especially so in the southern portion of the county, where a rich corporation, not satisfied with ONE HUNDRED FORTY THOUSAND acres of land, but by a system of bulldozing and petty persecutions are endeavoring to keep out honest settlers as well as to "freeze out" those who have lived upon their lands for years. . . . Men who have protected their homes when surrounded by hostile Indians will not tamely submit to be driven out by arrogant wealth.[17]

Such rhetoric only increased tensions, and both sides became more and more polarized.

Personal angers mixed with political tensions on both sides. After Hugh Glenn's death in 1883, his heirs removed financial control of the ranch from French, making him little more than an employee. Cattle prices declined as demands for money from the Glenn heirs increased, intensifying the pressures on French and making him less willing to compromise with squatters on company land.

Ed Oliver, a man who had worked on French's hay crews, began homesteading near French and petitioned the county courts for road access through the French-Glenn Company land. When access was denied, certain that French was responsible for the court's decision, he began meeting with the settlers who wanted French destroyed. Oliver had a reputation for violence, having been arrested in 1894 for beating a man nearly to death with a shovel.

On December 26, 1897, at about two o'clock in the afternoon, Oliver rode onto French's land. French and his crew were rounding up cattle and looked up to see Oliver galloping toward them. Oliver's horse struck French's so hard that the horse fell to his knees, and French struck out with his whip, beating Oliver about the head and shoulders. Oliver pulled out a gun and began waving it about, whereupon French turned his back on the man and rode off. And then, in front of all the crew, Oliver shot the unarmed French in the back, killing him instantly.[18] The jury—made up of homesteaders and shopkeepers in Burns— found Oliver not guilty, agreeing with his seemingly absurd claim that he acted in self-defense.

French's murder did not solve the problems between ranchers and squatters; it only ushered in an era of federal attempts to control the riparian boundaries that private ranchers had dominated for so long in the basin. Soon after French's death, a squatter named Sarah Marshall finally won a riparian case against the French-Glenn Livestock Company.[19] Ironically, the Marshall case established a precedent not for homesteaders, but for the public and federal nature of the contested riparian lands. The U.S. Supreme Court agreed that riparian rights were inapplicable, but then awarded the claim to Marshall as a claimant on the public domain. This decision gave the president a legal basis for declaring unclaimed lands around the lake federal property, thus enabling the establishment of a federal wildlife refuge on them seven years later. Although the homesteaders decried the monopolistic control of the cattle barons, their actions led to an era of federal control that was much more intrusive and, finally, much more devastating to their dreams.

RECLAMATION: REGIONAL AND NATIONAL EFFORTS

Frustrated settlers and homesteaders in the Malheur Lake Basin, as across the West, linked themselves with the growing national conservation movement— a key element of the Progressive Era. Reformers had dramatic dreams and hopes of what the control of water could accomplish in the region. Reclamation could turn the basin into a paradise of small, prosperous, truly American farms, what the editors of the *Burns Times-Herald* poetically called "bright visions of happy homes, of prosperous, contented people ... of countless smiling farms."[20] "LAND-LESS MEN FOR MANLESS LAND," trumpeted the *Harney Valley Items*, which described the basin as "an undeveloped empire, isolated and practically unknown" because of the "many thousands of acres tied up in large stock farms."[21]

By the 1890s many individuals were concerned that natural resources were being ruthlessly exploited, and they began to fight for regulated development of the public domain. A vision of scientific management, with its promise of efficiency, was at the center of progressive reforms. Progressive conservationists believed wholeheartedly that conservation meant controlling and thwarting corporate monopolists such as the cattle barons, and the way to do that was through developing resources for fair use. A federal reclamation law, reformers argued, would ensure the availability of resources for future generations and would allow for the scientific, efficient distribution of water. Reclamation became a rallying point for the progressive conservation movement.[22]

In the Malheur Lake Basin, as throughout the West, newspaper editors began to argue that it was the patriotic duty of the federal government to help enact this progressive vision of society. The editor of the *Harney Valley Items* wrote that

> Congress must soon take steps for national control of the whole irrigation problem. . . . Water must be economized that men may live. . . . There are in the pent-up hives of industry in the Eastern States and cities millions of men who, if the opportunity were afforded them through an opening of the West by irrigation, would swarm out of those hives and cover the western fertile plains and valleys with an intelligent and industrious population. . . . Under wise administration, Arid America has a glorious future. With her countless small farms and rural homes, communit[ies] where people live in the open air, till the soil with their hands . . . she will prove the sheet anchor of the Republic in times of national peril.[23]

These powerful claims helped lead to an increasingly strong federal presence in the West.

Federal support of reclamation had begun decades earlier, when Congress passed the Desert Land Act in 1877. Settlers could buy a 640–acre section of desert land, if they agreed to irrigate it within three years. Desert lands were defined as "all lands exclusive of timber lands and mineral lands which will not, without irrigation, produce some agricultural crop. The proof that the lands were desert depended upon the oath of two witnesses."[24] The hopeful settler paid twenty-five cents per acre on application, and another dollar per acre upon proof of compliance. This act, which marked the first major federal policy aimed specifically at reclamation of arid lands, proved vulnerable to exploitation by large economic interests. In the Malheur Lake Basin, as across the West, large cattle companies used the law to get control of riparian meadows bordering many miles of rivers and streams.[25] Residence was not required, so absentee investors were free to file on land and hold it for speculation.

Abuses under the Desert Lands Act renewed determination to reclaim the arid West, but as a public project, rather than one that relied on individual initiative. A severe drought in the mid-1880s helped convince legislators in Washington, D.C., that irrigation might be important for western settlement, with the result that Congress authorized irrigation surveys in 1888, 1890, and 1891.

In 1891 a Nebraska journalist named William Smythe attracted public attention with his articles on irrigation. He soon became the West's foremost reclamation advocate, developing a focus and ideology for the irrigation movement that transformed it into what the reclamation historian Michael Robinson calls a "broad-based popular movement to transform arid and semiarid lands into productive, small family farms."[26] Irrigation became perceived not just an economic reform, but as Smythe put it, "a philosophy, a religion, and a program of practical statesmanship rolled into one."[27] As Donald Pisani argues, "Irrigation promoters turned their quest for state and federal water projects into a religious rite."[28] Smythe's vision of irrigation was one of small farms, democratic values, and collective institutions. His most powerful argument was that the failure of irrigation under the Desert Lands Act proved that collective—rather than purely private—action was necessary for the orderly development of irrigated agriculture in arid lands.[29]

In 1894 Congress passed the Carey Act, designed to encourage cooperative state and private irrigation developments. The federal government would grant up to one million acres to each arid state on the condition that the state initiate the irrigation, settlement, and cultivation of those lands, disposing of them in tracts of 20 to 160 acres.[30]

After its acceptance by the State of Oregon in 1901, the Carey Act sparked a flurry of irrigation and drainage development in Harney County.[31] Within six years Oregon had already selected 432,203 acres of arid lands for reclamation.[32] But in Harney County, none of the Carey Act irrigation projects was completed, and by 1922 no Carey Act land was irrigated and producing a crop.[33] Projects failed for biophysical and social reasons: insufficient capital, poor engineering, alkali soils, bad drainage, short growing season, little knowledge of stream flow, poor markets for crops, and no transportation to those markets. The speculators who promoted projects rarely had sufficient capital or technical resources to carry them through completion. Construction techniques were shoddy. In Oregon water rights were still in such chaos that engineers and investors could rarely anticipate the amount of stream flow legally available. Nevertheless, promoters did not hesitate to overestimate the amount of available water, and then they radically underestimated the costs of transportation to markets. The result was, not surprisingly, failure.[34]

The disappointments of the Carey Act projects helped spur a push for federal involvement in reclamation, which was to have profound effects on southeastern Oregon wetlands and dry lands.[35] On June 13, 1902, the Federal Reclamation Act (the Newlands Act) passed, and President Teddy Roosevelt signed it four days later.[36] This act ushered in a new era in water control, helping to launch a progressive program that employed engineering as a tool for social progress. Under the terms of the Reclamation Act, public lands would be sold

1. The Blitzen River begins high on the slopes of Steens Mountain, nearly ten thousand feet in elevation. In this 1957 photo, snows linger well into July. (U.S.D.I. Fish and Wildlife Service, Malheur National Wildlife Refuge)

2. Snows from Steens Mountain feed the Blitzen River, which provides much of the water for the wetlands and lakes that make up Malheur National Wildlife Refuge. The west side of Malheur Lake, shown here, illustrates the wetland environment, with sections of open water alternating with emergent vegetation. (U.S.D.I. Fish and Wildlife Service, Malheur National Wildlife Refuge, R1–981)

3. *(Facing page)* By the time Malheur Migratory Wildfowl Refuge was established, managers faced the daunting task of trying to restore an ecosystem that looked as if it had nearly been annihilated. Decades of engineering projects helped create a huge riparian marsh complex in southeast Oregon—one large enough to cover Massachusetts, Connecticut, and Rhode Island. Surrounded by desert, these wetlands teem with avian diversity and abundance. *Above:* A colony of white pelicans nest near Cole Island Dike on Malheur Lake in June 1959. *Below:* Pelicans belonging to a flock of 1,850 use the lake in August of that year. (U.S.D.I. Fish and Wildlife Service, Malheur National Wildlife Refuge, *Quarterly Narrative Report, May–August, 1959*)

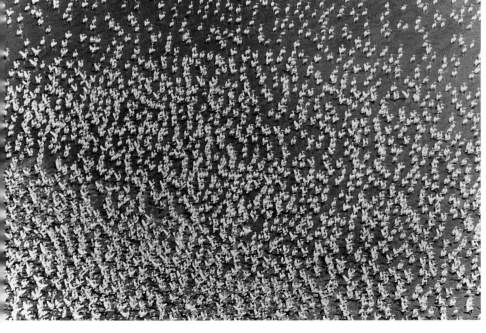

1 Wild as the Malheur Lake Basin seems, it has been radically transformed by ranchers, irrigators, and wildlife managers. The harvest of native grasses, the grazing of cattle, and the flood irrigation of meadows all altered riparian function. This native meadow hay is stacked for winter feeding, probably in the late nineteenth century. Sod House Ranch, to the south of Malheur Lake, is in the background. (U.S.D.I. Fish and Wildlife Service. Malheur National Wildlife Refuge)

5. Growing conflicts between homesteaders and ranchers over the control of riparian boundaries led to the murder of Peter French in 1897. *Above:* French in 1892, in a rare surviving photograph. *Below:* In 1935, Tebo was the last of the six Mexican vaqueros who had accompanied French into the Blitzen Valley in 1872. (U.S.D.I. Fish and Wildlife Service, Malheur National Wildlife Refuge)

6. Conservationists won a major victory in 1934 when the failed cattle and irrigation empires along the Blitzen River were sold to the federal wildlife refuge system, beginning the expansion of an empire of ducks at Malheur. The Oregon ornithologist, photographer, and filmmaker William Finley led the campaign to save Malheur. Here Finley is photographing birds at Malheur Lake in 1918. (Photo courtesy Oregon Historical Society, Finley Collection, negative D16)

7. On one of Finley's earliest visits to Malheur Lake, he stumbled on a colony of egrets. The adults had been slaughtered by plume hunters, and the young were left to starve slowly to death. Finley's horror at the scene motivated him to begin a campaign to save the great marshes of southeastern Oregon. Here Finley is searching for egrets on Malheur Lake in 1908; out of hundreds of thousands that had once nested there, only 121 were left by this time. (Photo courtesy Oregon Historical Society, Finley Collection, negative 2311)

8. By the early 1930s, after years of drought and irrigation withdrawals, the open waters of Malheur Lake had shrunk to just a few hundred acres. As the lake receded, squatters moved onto the lake bed and shallow wetlands, claiming the newly exposed land for themselves. This squatter's shack, surrounded by water, features a canoe and boat landing. (U.S.D.I. Fish and Wildlife Service, Malheur National Wildlife Refuge)

9. Critical labor for the daunting task of engineering the refuge systems came from the Civilian Conservation Corps. Here, ccc workers dig post holes for a fence along the refuge boundary in 1936. (U.S.D.I. Fish and Wildlife Service, Malheur National Wildlife Refuge, *Quarterly Narrative Report* September–November 1936)

10. In response to what they saw as a crisis, refuge managers in the 1930s adopted drastic measures to reflood drained lands, reroute water courses, and essentially manufacture new breeding areas for bird populations that seemed on the verge of extinction. With the help of the Civilian Conservation Corps, refuge staff bulldozed ponds, built dams, dug ditches, and extended hundreds of miles of canals along the valley. As shown here in 1937, the canals were designed for the efficient movement of water to refuge projects. (U.S.D.I. Fish and Wildlife Service, Malheur National Wildlife Refuge, *Narrative Report,* Camp Buena Vista, April 1937)

11. *(Facing page)* Managers' initial interventions seemed little short of miraculous. *Above:* Canals and ditches for water control near Buena Vista on the refuge are under construction in March 1937, and the landscape is dry and barren. *Below:* Just seven weeks later, the landscape has been radically transformed: the canals are finished, and water has been delivered across the valley. Within weeks, waterfowl were using these new wetlands. (U.S.D.I. Fish and Wildlife Service, Malheur National Wildlife Refuge, *Narrative Report,* Camp Buena Vista, March–April 1937)

12. These newly constructed wetlands are featured in the refuge's annual report in 1937. The caption reads, "Refuge Developments. Man-made Water Areas where Waterfowl Romp." (Scharff, "Report of Activities Fiscal Year 1937, Malheur Migratory Waterfowl Refuge")

13. The labor involved in constructing these "man-made water areas where waterfowl romp" was often extraordinary. Here CCC workers build a temporary manure dam in June 1937 so that water can be diverted from the main watercourse and spread over surrounding fields. Peter French's workers had used similar technologies to move water across their meadows in earlier decades. (U.S.D.I. Fish and Wildlife Service, Malheur National Wildlife Refuge, *Narrative Report,* Camp Buena Vista, June 1937)

14. Maintaining water-control systems was a never-ending task. When beaver and musk-rats returned to the valley, the refuge had locals trap them out, since the rodents' bur-rowing could destroy dikes. This man is skinning muskrats trapped on the refuge in 1943. (U.S.D.I. Fish and Wildlife Service, Malheur National Wildlife Refuge)

15. Even with the trapping program, the dikes and canals needed constant work. This dragline is working on a dike along the central canal on Malheur Refuge in 1949. (U.S.D.I. Fish and Wildlife Service, Malheur National Wildlife Refuge)

16. Funding for much of the work on the refuge came from an extensive and contro-versial grazing program. *Above:* Cattle graze on depleted refuge land in July 1937. The original caption, written by refuge manager John Scharff, reads, "A Source of Revenue for Years to Come." (Scharff, "Report of Activities Fiscal Year 1937 Malheur Migratory Waterfowl Refuge"). *Below:* Much healthier-looking cattle roam the refuge in 1956. (U.S.D.I. Fish and Wildlife Service, Malheur National Wildlife Refuge)

17. Control of willows was for decades a major chore for refuge staff. This staff member applies 2,4–D to willows growing along the Buena Vista Dike on Malheur Refuge in 1949. Note the lack of protective gear. Although the water is hidden by the willows, this spraying is right along the watercourse. (Photo no. 1400 by Ray C. Erickson; U.S.D.I. Fish and Wildlife Service, Malheur National Wildlife Refuge)

18. Carp presented a major challenge for refuge managers, since they destroyed habitats that waterfowl depended upon. As carp populations increased in Malheur Lake during the 1940s and 1950s, duck populations dropped. Beginning in 1955, carp control became a massive program for Malheur Refuge. These men are loading a spray plane with drums of rotenone, a potent fish poison, to be sprayed over Malheur Lake in an (unsuccessful) attempt to rid the lake of carp. (Photo no. 55–68 by David Marshall; U.S.D.I. Fish and Wildlife Service, Malheur National Wildlife Refuge)

19. *(Facing page)* The spraying of poison killed millions of carp, but enough survived to rapidly repopulate the lake. These are only a tiny fraction of the carp who died after the spraying in 1955. (Photos no. 55–71 and 55–87 by David Marshall; U.S.D.I. Fish and Wildlife Service, Malheur National Wildlife Refuge)

20. *(Facing page)* Part of the difficulty in killing carp lay in the complexity of riparian habitat. *Above:* The Silvies River, shown here in 1955, provided numerous channels and backwaters in which carp could hide. *Below:* Men struggle to paddle their canoe through the woody vegetation along the Silvies, trying to spray rotenone to kill carp during the 1955 program. The man in the bow of the canoe is operating the sprayer. (Photos nos. 55–91 and 55–101 by David Marshall, U.S.D.I. Fish and Wildlife Service, Malheur National Wildlife Refuge)

21. John Scharff managed Malheur National Wildlife Refuge for decades, gaining a great deal of political influence during those years. Here Scharff (in the middle) visits with Supreme Court Justice William O. Douglas and a friend on Steens Mountain in 1948. (Photo no. 1410 by Ray C. Erickson; U.S.D.I. Fish and Wildlife Service, Malheur National Wildlife Refuge)

22. Biologists at the refuge work to understand the dynamics of waterfowl populations and wetland habitats. (U.S.D.I. Fish and Wildlife Service, Malheur National Wildlife Refuge)

23. Ducks weren't the only birds of concern at Malheur. This long-billed curlew settles over eggs on the northwest side of Malheur Lake in 1948. (Photo no. 1411 by Ray C. Erickson; U.S.D.I. Fish and Wildlife Service, Malheur National Wildlife Refuge)

24. Sandhill cranes dance during the spring mating season in 1957. (Photo no. 57–7BW-12 by David Marshall; U.S.D.I. Fish and Wildlife Service, Malheur National Wildlife Refuge)

and the money would go into the Reclamation Fund, which would build and maintain dams, reservoirs, and canals under government management and control. The goal was to reclaim western lands and encourage settlers, not speculators. Individuals would then buy the resultant water, paying fees that would go back into the fund to pay for future projects.

The Reclamation Act was designed to foster small farmers and promote agrarian communities. Government aid was intended to counter monopolies, not create them. Nevertheless, agribusiness, not small family farmers, soon gained control of most of the irrigated acreage across the West.[37] In *Rivers of Empire*, Donald Worster argues that federal water programs led to a "coercive, monolithic, and hierarchical system ruled by a power elite based on the ownership of capital and expertise."[38] Agribusiness, the Reclamation Service, and banks together formed this hydraulic elite, manipulating nature, individuals, and communities. As Worster writes,

> Working together rather than in opposition, capital and bureaucracy have drastically reordered the arid region. They have shared a common logic, a broad plan of conquest, in the name of economic growth. They have achieved, as intended, an "empire" that today boasts forty million irrigated acres and sprawling metropolises. They have accumulated power as well, just as [John Wesley] Powell feared they would.[39]

Many historians object to Worster's characterizations of a hydraulic elite. Norris Hundley, for example, argues in *The Great Thirst* that although governmental bureaucracies and private business interests were indeed closely allied, they never formed a cohesive power elite. His studies of California water development show "the activities of a wide and often confused and crosscutting range of interest groups and bureaucrats, both public and private, who accomplish what they do as a result of shifting alliances and despite frequent disputes among themselves." Although Hundley agrees that reclamation did concentrate power in the West, he argues that competition among water users was important in shaping access to water and power in the region.[40] Pisani has shown that Westerners, always uneasy about government concentrations of power, were more supportive of the marketplace than of government planning.[41] Although the federal government built a massive infrastructure for water projects, it failed to settle questions of distribution, allowing local interests to determine access to water, often leading to inequitable concentrations of power in the hands of a few.

The Reclamation Act raised powerful hopes—however unsustainable they turned out to be—in the Malheur Lake Basin while stirring a complex pot of rage against cattle barons, longing for paradise, and dreams of quick riches. The *Burns Times-Herald* predicted that the act would result in "an inland sea blooming with industrious farms," and the *Harney Valley Items* noted, "The creation of an arid land reclamation fund . . . is a proposition which will commend itself to every interest."[42]

OREGON WATER LAW

The Reclamation Act rested on a progressive vision of scientific efficiency. Engineers, scientific experts, and planners would manage the funds, design the projects, and assure success. As a first step toward efficiency, the Reclamation Service refused to develop water projects unless states could guarantee a sound structure of water rights. The Reclamation Service thus became a powerful advocate for the creation of a workable system of water laws.[43]

In the wake of the Reclamation Act, Oregon designed a water law system in 1909 that combined riparian law and prior appropriation.[44] As early as 1864, Oregon had legislated that miners could claim water rights, and in 1868 the legislature allowed landowners to apply for permits for draining wetlands and modifying stream channels. In 1891 the state defined the first beneficial uses: irrigation, livestock, and domestic uses (with mining and hydropower added in 1899). Yet, because water rights throughout the state were still in chaos in the early 1900s, business and government leaders lobbied the 1909 legislature to pass a law declaring water a public resource and requiring a permit for anyone to use it.

The resultant Oregon Water Code of 1909 established four general principles: (1) water would belong to the public, (2) rights were for use only and would be assigned by the state through a permit system, (3) water use would follow the doctrine of prior appropriation, and (4) permits would be issued only for beneficial uses that involved the physical control of water, and only for uses without waste (although what constituted waste was not defined in the code). Finally, the law included a court-based process for settling disputes over water rights.[45]

The requirement that water rights must be for beneficial uses was intended to resolve conflict. Yet it only increased conflict, for it assumed society could agree on which uses were beneficial and which were not. Different social groups had radically different definitions of what constituted beneficial use, and those definitions changed over time.[46] Unless forfeited by lack of use, water rights lasted forever, which made the task of adjusting water use in response to changing conceptions of beneficial use extremely difficult.

In 1909, under the reformed water laws of Oregon, engineer John Lewis was named the new state engineer of Oregon. His task was to guide the development of reclamation in the state. Lewis responded with great enthusiasm to the new water code, declaring that under the old law, "utter confusion prevailed as to the legal status of a water right" and the result of such confusion was that "capital declined to invest" in the state. Lewis declared that the new law would prove to be of "as great importance to Oregon as was the making of the 'Doomsday Book' in 1085, by William the Conqueror, which was the first to attempt in England to systematize land titles."[47] But the state engineer was too optimistic: Oregon water law rested on the concept of beneficial use, and coming to a consensus about what that constituted proved impossible.

RHETORIC OF NATURE AND CULTURE

John Lewis began his job as state engineer, curiously enough, not by engineering anything but by writing a history of irrigation in Oregon. His intention was to help the state come to some consensus on beneficial use, and to do so, he had to convince ranchers—or at least the courts—that beneficial use meant ending flood irrigation and natural floods and replacing them with the modern machinery of reclamation: storage reservoirs, drainage, and efficient irrigation systems. Lewis's history portrayed the development of efficient irrigation and drainage methods as part of a "natural" evolutionary progress of development—so natural, in fact, that no one could rightfully question or oppose it.

Rather than setting irrigation in opposition to nature, Lewis claimed that irrigation itself was a form of nature. The very first irrigation systems, Lewis argued, were the spring floods that each year overflowed river banks and irrigated the surrounding flood plains, creating a rich network of riparian meadows.[48] In his vision of progress, these natural floods were the first step in an orderly, organic development toward reservoirs and storage systems. The next step in this evolution was flood irrigation. Lewis painted a portrait of a remote, simple past where flood irrigation had once made sense:

> In those early days, there was ample water in the streams for the use of all, even with the crude and wasteful use then prevailing. . . . To confine it in ditches, to spread it regularly over the land, to treat it as something precious, would cause an increased expense, and an unnecessary expense, at that time. To turn the water loose upon the land, let it run as it would without regulation, helped make pasture and make hay on the meadows.[49]

For Lewis the engineer, flood irrigation was a primitive art, for ranchers merely let the water run as it would "without regulation," without science, and without order.

The problem, in Lewis's eyes, was that by 1909 the natural evolution from flood irrigation to storage reservoirs had been stalled. He wrote, "In Harney County irrigation development is in its first stage, about 90 percent of the 80,000 acres irrigated being watered by the natural overflow of streams aided by primitive diversion works."[50] Rather than being part of nature, as they claimed themselves to be, ranchers were impeding nature by preventing the evolutionary progress toward reclamation. Lewis wrote,

> To do as was done in Harney valley in the olden time and spread the flood waters over the lands, to divert the waters of the streams and permit it to run over the lands and thus increase the natural hay crop or the pasturage, might be a beneficial use of the water under certain conditions, but when the county came to its own and all of the water was required in order to cultivate all of the lands, it became a wasteful use.[51]

Lewis urged that in a modern era when "railroads pierce the country in different directions and a market exists for soil products," it was man's moral duty to regulate the boundaries between water and land as efficiently as a machine. Conservation meant full use, and that meant storage reservoirs, drainage, and irrigation. He concluded, "What the Almighty had provided for making a rich, populous section out of Harney Valley, must be used to the fullest extent. . . . Wise public policy makes such conservation and use a necessity."[52] In the eyes of Lewis and other irrigation engineers, the flood irrigation practiced by ranchers was "primitive" and therefore needed to yield to what developers saw as the "natural" imperative of industrial progress.

The struggle to control the ways riparian nature would be reshaped was a struggle over the control of the rhetoric of nature, the control of people, and the control of water itself. Ranchers and reclamationists each saw the other's use of water as profoundly wrong and wasteful. Ranchers argued that allowing riparian areas to remain intact was a beneficial use, for they could profit from meadows without extracting water from them. Yet in arguing this, their motivations stemmed from a desire to hold onto water rights, not from any great love of intact riparian areas or undisturbed nature. Even while ranchers were turning more and more to mechanized operations, they increasingly used the rhetoric of nature to argue that they were the true progressive conservationists.

Irrigation reformers and settlers, on the other hand, also used the rhetoric of nature to support their own positions. An essay on reclamation written in the 1920s by the radical social critic Anne Martin illustrated the ways rhetoric could be used by those who viewed reclamation as a tool for social justice. Martin described the Great Basin as a "vast, exploited, undeveloped [region] with a meager and boss-ridden population. . . . The fundamental cause of every one of these conditions undoubtably [sic] lies in the monopoly by the livestock industry of the water, the watered lands, and the public range lands." Such monopoly had prevented the development of small farms and stable families, creating a migratory army of single, homeless, hopeless men. If it were not for the "stranglehold of the livestock industry," Martin wrote, the Great Basin could be irrigated to create small farms for diversified and intensive agriculture, which would allow true communities, with stable homes, women and children, and schools and libraries.[53]

For Martin, the solution was federal reclamation: irrigation projects that would support small farms and lead to the subdivision of corporate cattle holdings. Although she saw the solution as the separation of water from the land, she considered the result not to be an unnatural conquest of nature but the creation of a better nature. "Storage and distribution of water" would not destroy nature but would make it "blossom as the rose." For Martin, as for other irrigation reformers, riparian areas were sites of natural waste, not natural abundance. She wrote that she had

seen rivers flooding their banks on their way through barren valleys which in the language of congressmen would "blossom as the rose" with the storage and distribution of this water. The Humboldt River spreads out into a lake at one point, owing to a bad channel, and loses 300,000 acre feet in a few miles, due to evaporation and absorption. This is enough to irrigate 200,000 acres and provide homes for 2,000 families.[54]

In Martin's eyes, through their use of flood irrigation, ranchers had aligned themselves with natural waste rather than with natural abundance, and the result was human suffering:

I have seen large quantities of [water] overflowing the ditches and running to waste on the fields and roads of company ranchers, producing a rich crop of willows and tules after irrigating the wild hay lands . . . [while] across the road were the scattered 'dug-outs' and cabins of settlers . . . struggling to 'prove up' and sustain life for their families and themselves on a 'dry' farm. . . . Staring at us through the sage brush . . . were two or three eerie little children, timid as jack-rabbits, growing up without school or toys.[55]

Waste abounded, in Martin's view, as rivers overflowed their banks, and together with the ranchers and the courts, such waste denied the poor the right to their dreams.

Although many people today assume that machines and nature exist in opposition, reformers imagined a new world where nature and machines would join together, creating a better nature and better machines. As William Cronon notes, to the reformers, waste lands "all existed in a state of nature that prevented their exploitation until human ingenuity could 'reclaim' their potential—a potential that people tended to see as 'natural' even though it served human desires and cultural values far more than the needs of existing ecosystems."[56] For example, in one Harney Basin court case over water rights, the lawyer for the developer used familiar biblical rhetoric to paint a picture of a fallen nature waiting to be redeemed, not destroyed, by the engineer:

> Such a dream had McConnell during all the weary years that have passed since he first entered upon his endeavors to devote all the waters flowing through Harney Valley to their highest beneficial uses. He has dreamed that by so doing, he will be instrumental in establishing the family, the home and the school, where now only the sagebrush grows, and the stagnant pools abound; he had dreamed that through his efforts the desert and waste places will be reclaimed and made subservient to the will of man; that the barren plains comprising such large areas of the great Valley, and which now lie almost useless and worthless in the light of the western sun will be transformed into fertile fields and grassy meadows. . . .[Reclamation] will make out of that Valley a veritable Garden of Eden, an earthly paradise, its surrounding peaks standing at eternal sentinel guarding the beautiful expands of meadow, garden and fertile field, which ten fold the number of people now residing there will proudly call their homes.[57]

HANLEY AND IRRIGATION DEVELOPMENTS

As ranchers struggled to protect their water rights from reclamationists and settlers, those same ranchers began speculating in land by transforming their riparian meadows with dredging and irrigation, and becoming developers as much as ranchers. Yet the more they transformed the landscape into an agricultural machine, the more their rhetoric allied them with what was natural and therefore good. Although both ranchers and reclamationists manipulated the rhetoric of nature in their battles against each other, their work upon the land began to become increasingly indistinguishable, as both began to strive for more efficient control of water. The resulting landscape was what the historian Mark Fiege calls a hybrid landscape, a mixture of natural and artificial.[58] The social result was also a hybrid, as ranchers transformed into developers, farmers, and politicians.

The social historian Peter Simpson argues that it was the modern businessman who really won the classic feud between homesteaders and cattle barons by combining ranching, speculating, and irrigating.[59] The complicated career of William Hanley—Harney County rancher, developer, and politician—best illustrates this phenomenon.

Born in 1861 west of the Cascades, Bill Hanley left his father's mule ranch as a teenager and drove a small herd of cattle over the mountains to join his older brother John. Another brother, Ed, soon followed.[60] The three brothers, hoping to get rich quickly, started buying up land in Harney County from failed homesteaders who had come with the same high hopes. Ed and John soon abandoned the dream, but Bill stayed, buying up more land and cattle, finding beef markets through family connections in Alaska, developing deals, and slowly getting rich.[61] By 1912 he owned seven thousand head of cattle, and he drove them to the railroad four times a year, his wife, Clara, going on ahead in a buggy to prepare the way by setting up camp, finding water sources, checking fences, and smoothing relations with neighboring settlers.

After Peter French was murdered in 1897, the new owners of what had been the French-Glenn operation asked Bill Hanley to manage the company's lands. Soon Hanley controlled 195,163 acres of private land, nearly a third of the county's private acreage. Like Peter French and Henry Miller, Hanley steadily consolidated his holdings until, by 1912, as L. Alva Lewis, an agent for the U.S. Biological Survey (which later became the Fish and Wildlife Service), wrote of Hanley's operation, "The major portion of the good agricultural land of the county is theirs. . . . They own the valley."[62]

Like those of the earlier cattle barons, many of Hanley's holdings came from swamp land frauds.[63] Yet, although he was a rancher, he was also skilled at manipulating local hatred of cattle barons to support his own ambitions. Hanley became the major speculator in the county through his corporation, the Harney Valley Improvement Company.[64] He did many things the cattle barons would never have done: he encouraged settlement, tried to develop irrigation projects, fought for a railroad, and led the battle for paved highways and good roads.[65] He even ran for senator and governor, cultivating the friendship of national figures including Theodore Roosevelt, William Howard Taft, William Jennings Bryan, and James J. Hill. As the *Burns Times-Herald* bragged, "When Will Rogers stopped here between planes on his fatal flight into the northland last month, the first person he inquired of was 'my old friend, Bill Hanley.'"[66]

Unlike the cattle barons, Hanley was in many ways a town man with town interests and a town home. In the local historical museum in Burns, an entire room is devoted to him. The room creates an image not of wild cowboy days and the mythic West, but of middle America striving to succeed. With its velvet armchairs, leather-bound books, pictures of dead animals, and a table covered with papers, it evokes the life of the family man, the good citizen, the politician, the businessman—not the Wild West life of a cattle baron. Even though Hanley was a very rich man—far richer than Peter French ever became—he had a populist suspicion of powerful finance and a fear of corporate dominance, and when he ran for U.S. senator in 1914, it was on the Progressive ticket.[67]

In 1901, right after the Oregon State Legislature passed a law accepting the Carey Act for irrigation, Hanley formed the Harney Valley Improvement Company to reclaim the arid lands that were still vacant in the valley. In partnership with Drake O'Reilly, an investor from Portland, Hanley designed a series of dikes, ditches, and reservoirs to divert the flow of the Silvies and drain marshes in hopes of irrigating nearly one hundred thousand acres of arid land in Harney Valley. Half this land was federal, and Hanley hoped the government would grant it to the state after the improvements. Twenty thousand acres belonged to Hanley and ten thousand acres to the Willamette Valley and Cascade Mountain Wagon Road Company. Within a few years, Hanley had convinced investors to put $100,000 into the project. He figured that he would profit on the contract, improve the value of his own land, and control eighty thousand acres of grazing privilege for free while the work was completed.[68] The only restriction was that Hanley could not start on canal construction until he had a contract with the federal government under the Carey Act, and there the project stalled.

Twenty thousand acres of wetlands that Hanley hoped to reclaim belonged to Henry Miller's Pacific Livestock Company. On October 12, 1901, the *Harney Valley Items* reported,

> This project, however, will arouse the hostility of those who profit from the present condition of the country, and it is not improbable that the courts will be asked to decide where the equities and right lie. Stockmen own the marshland, which produces heavy hay and is excellent pasture. . . . They do not want the country to become thickly settled and tilled for general crops, for that would restrict the open range. . . . The Valley would be rendered more productive and would give homes and employment to a large number of people whereas it is largely unsettled now. . . . It is estimated that the water to be obtained from the Silvies is sufficient to irrigate all the land in the valley lying under it, but this must operate to dry up the extensive marshes about the lakes and the marshes will then become plow land.

As the *Items* had predicted, Miller's Pacific Livestock Company objected to the project because the company would lose winter pasture at the mouth of the Silvies River, and those marshlands were more valuable to them wet than dry.[69]

Hanley negotiated a contract with the state to support his irrigation project and then went to Washington, D.C., to meet with Mr. Newell of the new Reclamation Service, hoping to have his project developed. Raising the hopes of local irrigators and homesteaders, in June 1903 the secretary of the interior ordered the temporary withdrawal of 1,080,000 acres of land in Harney County, along the Silvies and Malheur Rivers. The Harney project alone included, at that point, 622,000 acres.[70] For two more years, the state negotiated with the federal government. In 1903 developers also proposed a project on Silver Creek, where there were fewer conflicting water rights and claims. The *Harney Valley Items* reported, "It is stated on good authority that the first work of government reclamation in this section will be commenced in Silver Creek valley. No hindrances lie in the way of the government in regard to prior claims to the territory."[71]

Although locals were enthusiastic about both development projects, government engineers were more cautious. The report of one consulting engineer stated,

> While the present method of irrigation by flooding is practiced, in which from two to five times the amount of water is used than would be necessary if it were properly distributed, it is doubtful if in seasons of average precipitation there is more water than is required to satisfy present claims. In June, 1904, the board of consulting engineers examined the project and recommended that it be abandoned, since the water supply would be found inadequate.[72]

Finally, in 1906, Mr. Maybury, a special representative of the secretary of the interior, ruled against both proposed reclamation projects, saying that the lands were not arid and did not require irrigation. As one lawyer put it, Maybury objected to the "non-desert character of a portion of the lands, and the non-feasibility of the plan and method of irrigation, which questioned the water supply for so large an acreage and the proposed method of using the flood waters during their natural flow, without the added element of storage."[73]

When the Reclamation Service ruled against the projects, local newspapers railed against the federal engineers, the railroad barons who were not developing the region, and the cattle barons who retarded progress simply by existing.[74] Hanley had once been viewed as the noble developer of the region, but suspicions and resentments soon built against him. He was called the unscrupulous head of the "Carey land-grab concern" and many in the valley believed he had never intended to irrigate, only to speculate.[75] Lawyers opposing Hanley claimed in court that "the sole business of the Harney Valley Improvement since 1909 has been dealing and speculating lands, and that it never had improved, by irrigation or otherwise, any lands held by it from time to time."[76]

When Maybury ruled against irrigation development, Hanley never reimbursed the original investors. In court, lawyers inquired about the fate of the investors' $100,000, asking, "Had it been diverted, even if water never was, and used for other purposes?"[77] Hanley flippantly replied, "The money that we originally had to construct with was so disappointed that it went to pieces and the money was not available afterwards."[78]

Even after the Reclamation Service dropped the Silver Creek project, Hanley continued to pursue it privately, working quietly with John Whistler, the service's irrigation engineer. Whistler wrote to Hanley on April 2, 1910, "I have had some men from Seattle in who wish to take up the Silver Creek project. They apparently have the money and I told them I thought it would be advisable at least to take up with you first the matter of your 'OO' property (the 'banana belt') and the water right on the lower creek."[79] Meanwhile, Hanley was dabbling in colonization schemes on another of his ranches. Whistler wrote to Hanley on July 1, 1910,

> If you decide to sell the property without developing it first, let me know what you propose to try to get out of it. I may be able to send you a buyer. . . . I'm almost afraid to say what I think you can make out of the project if you take it up in this way and are prepared to keep up the work actively until you have sold the property, even to the point of actually colonizing part of it, if necessary.[80]

Hanley, for all his interest in land speculation and development, still retained a fascination with the workings of riparian meadows. For example, when one lawyer in a water rights case asked him, "What water do you really need to grow crops here, and when do you need it?" Hanley replied by telling of his experience with a meadow on his OO Ranch:

> Through the series of years we commenced our regular habit of digging canals on the ranch, and finally got absolute control of the water that flowed into those swamps. . . . It was so nice and convenient to the operating part of it that when [the cows] were taken off at haying time, the water was kept out of the swamps and then kept out later in the spring.

Hanley found that "absolute control of the water" was far more convenient than dealing with an unregulated marsh. But he soon noticed that "the crops commenced to decrease on these meadows." As an experiment, he tried restoring water back to the meadow after he had cut the hay from it: "The last two years I have adopted the system of rushing the hay off of that meadow and putting the water back onto them and then putting the water onto this grass as early in the spring as we can get the cattle out of the fields, say the first of April." To justify this odd practice, Hanley argued from evolutionary history, claiming that it made sense because "meadow that originates from the salt grass in my judgment and has been all of its period of its evolution constantly laying with this water on top of it. . . . In all probability that was the history of its evolution on these high lands, meadows we call them in this country, for a great long period of time." As the lawyer put it, for maximum profit Hanley "found it necessary to restore the natural condition to a large extent."[81]

Hanley's fascination with the workings of water in that natural world reveals not just contradictions in his own endeavors but also the tensions inherent in the changing nature of water control in the valley. He was equally passionate about irrigation, reclamation, and bird conservation. While draining marshes and fighting federal proposals to establish a wildlife refuge in the basin, he was equally busy in his own efforts at marsh restoration. Even though Hanley fought the refuge bitterly, he considered himself a conservationist. He wanted to develop eastern Oregon but also to preserve elements of its wildness—the marshes, the birds, the great flocks of migratory wildfowl. He wanted to bring progress and paved streets and sanity to the region, yet he also wanted the frontier images of cowboys and Indians to stay intact.

William Hanley's largest project was management of the old P Ranch, Peter French's empire. French's death had ushered in an era of land speculation in the basin. In 1906 the French-Glenn Livestock Company was sold to Henry L. Corbett, twice acting-governor of Oregon, who hired Hanley to manage the properties for irrigation and drainage. Hanley organized the holdings into the Blitzen Valley Land Company, a firm of speculators. Hoping to subdivide the holdings and sell them off as small farms, Hanley built extensive irrigation facilities and dredged twenty miles of the river.[82]

The development scheme never paid off, however. In 1916 the operation reorganized yet again, this time into the Eastern Oregon Livestock Company, selling 40 percent of the stock to Louis Swift, the owner of a Chicago meat packing company. The Eastern Oregon Livestock Company managers ran about twenty thousand head of cattle but focused on land speculation and subdivision schemes. Unlike Peter French, the managers failed to put up enough winter hay, and they lost nearly half the cattle in the next two winters. Feral hogs filled the tule marshes and bottomlands, and then sheep were brought into the valley, which complicated conditions.

The effects of human, animal, and machine activity were apparent by the early 1930s. After four decades of overgrazing, irrigation withdrawals, grain agriculture, dredging, and channelization, followed by several years of drought, the valley had become a dust bowl. Attempts to increase production by making wet lands drier and dry lands wetter had stripped the willows and cottonwoods from the banks, imprisoned the river in a channelized ditch, and dried up the meadows and marshes. People didn't fare much better than the land. Ranches failed, livestock starved, homesteaders went bust, and the primary occupation in the valley became suing neighbors over water rights. Water control was an unmitigated disaster. French's P Ranch was sold once again, this time to the federal government, which wanted the Blitzen River water rights to protect water levels on the Malheur Lake Bird Reservation.

For many environmentalists who work on riparian issues today, reclamation efforts across the West represent human engineering of nature at its very worst. Many conservationists of the early twentieth century—the people who tried to save Malheur for birds—also believed that reclamation was a tool to destroy landscapes and defraud workers. For example, William Finley, the Oregon conservationist who almost singlehandedly saved Malheur from drainage, portrayed reclamation as a scheme led by evil men—promoters, railroad speculators, schemers in chambers of commerce—in league to destroy good land and good families. In Finley's vision, reclamation was driven by "the coming of the land promoter. He was the canker in nature's balanced system, a schemer who persuaded state and county governments that vacant desert lands could be turned into prosperous farms, that ponds and marshes could be drained and add agricultural wealth to the communities."[83] Yet Finley, like many environmentalists today, underestimated the power of the social vision that drove the reclamation dream. Reclamation was not just an attempt to drain the desert wetlands and irrigate the desert uplands; it was a social experiment that envisioned water resource development as the basis for a new civilization in the West. Controlling water by severing its connections with the land seemed to offer the chance to sever as well the bonds of social control that ranchers had held.

3 / Buying the Blitzen

B y the early decades of the twentieth century, increasing efforts at drainage and reclamation had led to a noticeable decline in ducks throughout the United States. These declines helped stimulate a national interest in conservation of waterfowl and the habitats that they depended upon, just while the conversion of wetland and riparian habitat to farmland was accelerating. This chapter examines a critical flashpoint in the conflicts over the transformation of riparian habitat in the West. After years of frustrating reversals, preservationists won a major victory in 1934, when the failed cattle and irrigation empire along the Blitzen River was sold to the federal wildlife refuge system, beginning the expansion of an empire of ducks at Malheur. This event signaled the growing power of a preservationist vision of riparian areas, a vision that was increasingly able to transform policies while influencing the transformation of landscapes as well. In their quest to control natural boundaries between water and land, preservationists, like ranchers and reclamationists, also struggled to control natural metaphors.

In 1904 and 1905 the Oregon biologist, photographer, filmmaker, and writer William Finley toured the great marshes of the southern part of the state. Finley was soon to become prominent in western wildlife conservation. Several years after this voyage, in 1911, he established Oregon's first Fish and Game Commission, and eventually he became state game warden, state biologist, and commissioner for fish and game. Finley had transformed his youthful passion for collecting birds into a love for photography, journalism, and conservation activism.[1]

As Finley paddled a little boat through the marshes of Malheur in the first years of the twentieth century, he found himself lost in a maze of marshes so trackless, vast, and confusing that he nearly persuaded himself he was the first person ever to ply their waters. Just as he was telling himself that Malheur was still an untouched Eden, Finley stumbled onto a scene of devastation that shocked him into action that would change his life: a colony of egrets slaughtered by plume hunters, the young left to starve slowly to death.[2]

Finley reacted to the site of ransacked colonies as if he had stumbled into the Garden of Eden just after Eve took a bite of the apple and passed it on to Adam. Paradise had been plundered, sullied with the stain of sin. Out of hundreds of thousands of egrets that had once nested in Malheur Lake, only 121 were left when Finley toured the region. His horror at the decimation motivated him to begin a campaign to save the great marshes of southeastern Oregon—a campaign that soon led to dramatic clashes with homesteaders, ranchers, and irrigation developers. On his return to Portland, Finley wrote feverishly, trying to publicize what he had found at Malheur before it was entirely diminished: the greatest concentration of ducks, shorebirds, egrets, herons, cranes, and ibises in the country, perhaps even the world.

Although Finley was a skilled ornithologist, his most powerful tool was not science but rhetoric. His task was to publicize the marshes of Malheur, and in the process to rouse public opinion within the state and across the nation in favor of their preservation. Yet this was no easy task, for reclamationists had already borrowed Edenic rhetoric for their task of redeeming the marshes from their watery grave. Finley had to subvert centuries of rhetoric that linked marshes with fallen nature and their drainage with redeemed nature. He had to convince a nation that drainage was destruction, not reclamation. To do this, he borrowed language from the reclamationists to create a new myth of Malheur that incorporated Edenic images with a particularly American myth of origins: that of the romantic cowboy.

Preservationists painted a portrait of Malheur as a place in the first days of creation, a place captured in the new light of dawn, when only the "red men" plied the waters. Alva Lewis, an inspector for the federal refuge system, wrote of Malheur in 1912,

> In Malheur it would appear that the Creator had exerted a special influence looking to the creation of a water fowl paradise. Almost every acre, even the open water of the lake has an abundance of vegetable life, while the tules of the marshes are rarely so dense as to prevent the growth of the various plant life on which water fowl feed. Tules, millfoil, pond-weed, duck weed, wocus [pond lily], goose grass, cattail, burreed, sugar grass, arrow plant, smart weed, wire grass, pepper mint, camas, water hemlock, and many other plants, the common names of which I am not familiar, can be found everywhere in abundance, I might say superabundance.[3]

For the conservationists, the cattle barons were part of this myth of Eden, as characters in a primitive drama, much like Indians. Lewis wrote,

On her immense stock ranches can still be seen the cowboy in his primitive glory, with the customs and methods of work of a half century ago. . . . 'Tis true there are cultivated areas—grain lands and tame grasses, but the farmer who tills the soil hardly counts. The stockman who pastures his cattle, horses and sheep on the public domain—who cuts the wild grasses of the natural meadows to feed his half-wild herds—he is the man who has made Harney county what it is today.[4]

This was extraordinary language for a government inspection report aimed not at the public but at fellow bureaucrats.

When conservationists wrote for the urban public, they evoked these Edenic images much more strongly. The work of Dallas Lore Sharp illustrates this well. Sharp, a close friend of William Finley, was a popular writer who did much to focus the national eye on the wild landscapes of Oregon. In the early decades of the twentieth century, he wrote about wilderness for an educated East Coast audience (his publisher was the Riverside Press in Cambridge, Mass.). A great sense of loss pervaded Sharp's writings, just as had Finley's essays. Both men felt as if they were witnessing a fall from paradise.

In *Where Rolls the Oregon* (1914) Sharp wrote of Malheur, "Here was a page out of the early history of our country." Once, all of America was an Eden, a place of unimaginable abundance:

> The accounts of bird-life in early American writings read to us now like the wildest of wild tales—the air black with flocks of red-winged blackbirds, the marshes white with feeding herons, the woods weighted with roosting pigeons. I have heard my mother tell of being out in a flock of passenger pigeons so vast that the sun was darkened, the birds flying so low that men knocked them down with sticks. As a child I once saw the Maurice River meadows white with egrets, and across the skies of the marshes farther down, unbroken lines of flocking blackbirds that touched opposite sides of the horizon.[5]

But in less than a generation, industrialization had destroyed this Eden. Malheur represented to Sharp all that had been lost throughout the nation:

> The sedges were full of birds, the waters were full of birds, the tules were full of birds, the skies were full of birds: avocets, stilts, willets, killdeers, coots, phalaropes, rails, tule wrens, yellow-headed black birds, black terns, Forster's terns, Caspian terns, pintail, mallard, cinnamon teal, canvas-back, redhead and ruddy ducks, Canada geese, night herons, great blue herons, Farallon cormorants, great white pelicans, great glossy ibises, California gulls, eared grebes, Western grebes—clouds

of them, acres of them, square miles—*one hundred and forty-three* square miles of them![6]

Sharp' history was one of an imagined American Eden, but no matter how unfactual, this was a history of great power, for its myths resonated with meaning for Americans who were witnessing rapid industrial transformation.

> For here in the marsh of burr reed and tule, the wild fowl breed as in former times when only the canoe of the Indian plied the lake's shallow waters, when only the wolf and the coyote prowled about its wide, sedgy shores. I saw the coyote still slinking through the sage and salt grass along its borders; I picked up the black obsidian arrowheads in the crusty sand on the edge of the sage plain; and in a canoe I slipped through the green-walled channels of the Blitzen River out into the sea of tule islands amid such a flapping, splashing, clacking, honking multitude as must have risen from the water when the red man's paddle first broke its even surface.[7]

For Sharp's urban audience, the remoteness of Malheur was a powerful trope:

> But it was the air, the aspect of things, rather, the sense of indescribable remoteness, withdrawal, and secrecy ever retreating before us, that seemed to take on the form as something watchful, suspicious, inherently wild, something wolf-like. This was the wildest stretch of land, the most alien, that I had ever seen.

In 1914 Sharp hoped that Malheur would be saved by this remoteness:

> Separated thus by the deserts from any close encroachment, saved to itself by its own vast size and undrainable, unusable bottoms, and guarded by its Federal warden and the scattered ranchers who begin to see its meaning, Lake Malheur Reservation must supply water-fowl enough to restock forever the whole Pacific slope.[8]

But he underestimated the developers. Rather than Malheur's being saved by its own vastness and remoteness, those qualities seemed to make it an even greater prize to speculators. Just a few months after Sharp's tribute to Malheur was published, a battle over the basin's riparian riches began that would drag on for twenty years.[9]

ESTABLISHING THE REFUGE

Finley and Sharp's writings about the glories of Malheur convinced the state Audubon Societies, and through them, President Roosevelt, that the marsh was a tremendous resource for the future of American wildlife. In 1908 Roosevelt established Malheur Lake Bird Reservation, which did not include the rivers that ran into the lake but only the lake itself. At the time few saw the riparian areas along the Blitzen and the Silvies Rivers as important for wildlife, and so no one tried to protect them. The lake was where the ducks were most visible, so the lake was what won protection. The riparian meadows that fed into the marshes, the creeks, and the slow-moving waters along the rivers seemed hardly worth worrying about at the time, for few biologists recognized that they might be critical for perpetuating the abundance of Malheur.[10]

Protection at Malheur had its origins in a national movement for wildlife conservation that had begun a generation earlier, largely stimulated by private efforts by scientists and birdwatchers. A growing interest in birds and nature study, linked with attention to the odd fashion of dead birds perched on ladies' hats, stimulated concern over declines in bird populations.[11] In 1886 the American Ornithologists Union estimated that in North America alone five million birds died for fashion. Whereas hunting was an obvious target for conservationists, habitat loss and its effects on wildlife began to emerge as a scientific concern soon after the turn of the century. Finley, for example, had been roused to action by the market hunting of egrets, but he soon realized that hunting alone was not the primary cause of bird declines. Other ornithologists followed Finley's lead as he turned from attacking hunters to enlisting their aid in the preservation of habitat.

Private efforts alone seemed inadequate to support the burgeoning conservation movement, and in 1892 President Benjamin Harrison set aside the first federal sanctuary specifically for wildlife: a national salmon-spawning reservation on Alaska's Afognak Island.[12] When Teddy Roosevelt became president in 1901, he began to create a network of federal refuges. The first was in Florida, on Pelican Island—a five-acre federally owned rookery for brown pelicans. Although President Roosevelt had the power to create refuges on federal land, the federal government had no clear power to spend money to manage them. Roosevelt's friend A. Chapman asked him to sell Pelican Island to the Audubon Society, which had the staff and money to protect the rookery. Fearing political trouble over the sale of federal property, Roosevelt instead issued an executive order on March 14, 1903, making Pelican Island "reserved and set apart for the use of the Department of Agriculture as a preserve and breeding ground for native birds."[13]

By the time Roosevelt left office in 1909, he had established fifty-three federal refuges. In the words of Ira Gabrielson, an Oregon ornithologist and eventually chief of the Fish and Wildlife Service, the year 1908 was "a banner one . . . [because] for the first time larger areas were reserved. Largely through the efforts of William L. Finley and a small band of supporters, Lower Klamath, Oregon, and Malheur Lake, Oregon, were set aside as nesting grounds for migratory waterfowl."[14] Congress, however, refused to appropriate money to manage the refuges, so state Audubon Societies hired wardens to protect the birds.[15]

Although Congress had not allocated funds for refuge management, lawmakers attempted in 1900 to protect birds by passing the Lacey Act, which prohibited interstate shipment of birds killed in violation of state law. But the law was rarely enforced, and proved ineffectual. In 1913 Congress enacted two statutes: the federal Tariff Act, which forbade the import of plumes and other bird parts except for scientific purposes, and the Weeks-McLean Act, which declared the protection of migratory game birds a federal responsibility. Knowing the bill would be challenged on constitutional grounds, environmentalists lobbied for a treaty with Canada to protect birds that crossed the border. President Woodrow Wilson signed the Migratory Bird Treaty in 1916, prohibiting the sale of game birds and giving the secretary of agriculture the authority to limit hunting seasons and impose bag limits. With this act, the federal government became the primary protector of waterfowl.

As historian Ann Vileisis points out, the Migratory Bird Treaty would become critical to the federal government's relation with wetlands, for treaty obligations held the federal government responsible for safeguarding wetlands as well as regulating market hunting. Concern over birds, therefore, sparked America's initial concern over the protection of wetlands. The result was that, for a few years, waterfowl made a comeback. Yet in spite of new refuges and new laws, waterfowl populations were not out of trouble. Within a decade, duck populations crashed as numbers of hunters increased dramatically. More important, waterfowl habitat was being destroyed at an astonishing pace, as drainage became "something of a national mania," in the words of a former chief of the Fish and Wildlife Service.[16]

THE ATTEMPT TO DRAIN THE LAKES

Drainage in the Malheur Lake Basin began to threaten Malheur Lake well before the refuge was established, and continued to threaten it for decades afterward. Efforts to drain the Blitzen Valley had begun in 1902.[17] The French-Glenn Livestock Company manager argued that the goal of drainage was not to speculate on land colonization schemes but merely to increase cattle carrying capacity "by at least 15,000 head." But company papers from that year stated that "in round numbers, 125,000 acres of these lands are agricultural and will grow barley, oats, rye and wheat as well as alfalfa, potatoes and vegetables," making it clear that the French-Glenn Livestock Company was already developing schemes for agricultural colonization.[18]

In 1903 the *East Oregon Herald* reported that the French-Glenn Livestock Company had begun to "reclaim the great swamps" along the Blitzen: "The intention at present is first to cut a channel for the river, and then construct laterals as occasion demands for the purpose of draining the area. When this is accomplished it will be necessary to devise a method of irrigating this drained land properly."[19] Yet this grand scheme soon fell apart because, in the words of a USGS geologist's report from 1909, "the yielding nature of the saturated peaty land soon allowed these channels to become choked up."[20]

When irrigation schemes failed to make the sagebrush desert blossom into small farms, William Hanley, like other promoters in the basin, turned to drainage of the marshlands and riparian meadows along the Blitzen and the lakes. Since the natural growth was so dense and luxuriant, developers believed that surely those lands would grow fine crops if only the excess water were drained off. As Ann Vileisis puts it, "Lawmakers often suggested that swamps simply covered the earth and could be easily removed like a blanket from a bed. . . . Rather than recognize that excess waters in swamps *supplied* bounteous riparian forests, vast flocks of waterfowl, and other natural riches, legislators, along with most citizens, thought that surplus water prevented lands from being even more abundant."[21]

After Senator Henry Corbett bought the Blitzen holdings in 1906 and made Hanley ranch manager, Hanley cleaned off the old dredger and turned it loose once again in the riparian meadows along the Blitzen River. In his quasiautobiography, *Feelin' Fine* (coauthored by Anne Shannon Monroe), Hanley recounted how much he admired the marshes he was digging up: he enjoyed seeing "the different layers of soil and the peat on top and the differences in vegetation." He admired the Indian arrow points that the dredger threw up, the volcanic points in the swamp, and even the rattlesnakes he hired a group of boys to kill. Above all, he admired the work his men were doing to dredge and drain the marshes. But he noted that their entry into that "great swamp" was an ambivalent victory, for "as soon as the canal was dug and the cow went in, the oldest known weeds began to spring up on the canal's banks, and flies and mosquitoes came a-plenty. Use seems to produce what we call enemies, but they may not be. They all have their purpose. We have lots yet to learn."[22]

Hanley's effort to channelize the Blitzen ground to a halt when the federal government ordered him to stop the dredging and served an infuriated Hanley with an indictment for "cutting juniper on the government's forest reserve" to run the dredger.[23] But by then, twenty-five miles of main drainage canal and ten miles of intercepting canals had been constructed, reclaiming nearly "20,000 to 25,000 acres of tule swamp," according to the calculations of federal engineers in 1916.[24]

Hanley developed a colonization scheme for the reclaimed land, creating the Blitzen Valley Land Company, which set out to sell the drained land to urban settlers. His company prospectus stated that the sagebrush desert

> offers the finest quality of land for settlers if water be supplied to it, and it still lies open for settlement to this day for the reason that all the water is in the control of the Blitzen properties. . . . A canal 25 to 30 feet wide, and 9 feet deep, has now been dredged through what has heretofore been tule and flag bottom from Rockford Lane to the river channel at the head of the Valley. This is rapidly reclaiming the overflowed lands and will at the same time serve as the main canal for irrigating ditches.[25]

In 1913 Oregon passed the Thompson Act, which authorized the drainage of lakes in Oregon by any person or corporation whose proposals were approved by the state Land Board. This act spurred a flurry of interest in drainage possibilities in the Malheur Lake Basin, which accelerated when the Oregon District Irrigation Law was passed in 1917, allowing irrigation districts to be formed and paid for with bond issues. Across the nation, it was not until the advancement of technology for stream dredging and tile drainage that vast expanses of wetlands could be transformed into farmlands. But as drainage became technologically feasible, state legislatures began to pass statutes authorizing the creation of cooperative drainage districts. These became, as Vileisis wrote, "the cornerstone institutions for successful large-scale conversion of swamplands." Drainage districts gained the authority to tax landowners to pay for drainage works—a radical concept that altered the nature of government and the face of the landscape.[26]

Drainage fever began to threaten not just the rivers supplying water for the federal wildlife refuge but soon the heart of Malheur Refuge itself. Just after the passage of the Thompson Act in 1913, a local businessman named W. C. Parrish applied to the state Land Board to drain Malheur Lake, filing an appeal for the withdrawal of lands bordering the lake.[27] He did not seem to be troubled by the fact that the lake was a federal bird refuge. By June of that year, the Burns lawyer C. B. McConnell, acting for the Harney Basin Reclamation Company, also applied "to drain and reclaim Malheur Lake, desiring the state to apply to the United States Government for the title to the lake-bed, under the Oregon Swamp-land Act."[28] Soon eight separate attempts to lay claim to federal refuge lands, with the purpose of draining them, were filed.[29]

Malheur Lake was located in one of the sites first surveyed for reclamation potential by the new Reclamation Service. The state engineer reported that "investigations were first begun in Oregon by the U.S. Reclamation Service in 1903. The Malheur River and Harney Valley projects at first appeared most promising, and detailed surveys were prosecuted with a view to early construction. These projects were abandoned on account of certain complications," namely, the contested water rights on the Silvies River. Henry Miller's Pacific Livestock Company had thwarted all projects that might have taken water from the Silvies' riparian meadows.[30] Ironically, what saved the Malheur Refuge from being destroyed by drainage along with the other federal refuges in the region were precisely its tangled water rights and the stubbornness of local ranchers.

Although Reclamation Service engineers had recommended against including Malheur Lake Basin sites, efforts to include the basin in a federal project had continued until a 1909 U.S. Geological Survey study showed that most reclaimed soils in the basin would be worthless for agriculture. The 1909 report, by the geologist Gerald A. Waring, noted "the high percentage of soluble salts and their persistence to a depth of at least 6 feet. . . . A harmful quantity of these salts is present; and as drainage conditions are poor the soil is almost worthless for crop production."[31] Although Waring warned of poor agricultural prospects, his report noted that squatters were moving onto the lakebed even as the refuge was being established. He wrote, "A few settlers have taken up claims on the borders of Malheur and Harney lakes. They were probably attracted to this section by the growth of salt grass and by the moist character of the land, but it is to be regretted that they were not informed of the worthlessness of such land for agriculture."[32]

By 1909 federal reclamation engineers had recommended against developing reclamation projects in Malheur Lake Basin; federal soil scientists had warned of the worthlessness of the lake bed for agriculture; the secretary of the interior had ruled against all project proposals for the basin; and the president of the United States had proclaimed the lake to be a federal wildlife refuge. Surely, it seemed to Finley and other conservationists, drainage would not proceed. But all these impediments to drainage seemed only to stimulate, rather than discourage, the drainage promoters—private, state, and even federal. By 1916, a full eight years after the founding of the refuge, the Reclamation Service and the State of Oregon were once again plunging ahead into drainage efforts, urging complicated projects to drain the lake, move Blitzen River water to the Silvies River valley north of the lake, resurrect the Harney Project, and turn the valley into a complicated machine for the drainage and delivery of water. For three more decades the federal refuge had to fight against these projects for its survival, and for most of that time few believed the refuge could have a hope of winning.

In 1916 the State of Oregon filed a claim for title to Malheur Lake so that drainage efforts could proceed. Oregon based its claims on its state ownership of navigable bodies of water.[33] The federal government challenged this claim to title, arguing that—as the U.S. Supreme Court had ruled in the Sarah Marshall case—Malheur Lake was not navigable. Nevertheless, while this suit wound its long way through the courts, the Reclamation Service joined forces and funds with the Oregon State engineer's office to write a report on draining Malheur Lake for agriculture.

John Whistler, federal reclamation engineer, and John Lewis, the state engineer, coauthored the 1916 joint report that recommended drainage of Malheur Refuge. One of the striking qualities of this report is the complete lack of attention paid by the engineers to the federal wildlife refuge that already occupied the lake they wanted to drain. Throughout the entire report, neither engineer mentioned the existence of the refuge as a possible impediment to the project. This is particularly remarkable, given that only a decade earlier the Reclamation Service had rejected two federal reclamation projects because of complicated water rights.

Whistler and Lewis ignored not only the fact of the federal refuge and its possible rights to water but also Gerald Waring's federal report on soils in the basin. They did acknowledge that alkalinity might conceivably present a problem for irrigation projects, yet they optimistically hoped that drainage would *fix* this.[34] In the Klamath Basin, near Malheur Lake Basin, engineers did figure out ways to construct deep drains that would use wastewater to leach alkali salts out of irrigated soils. But in most places the efforts were not as successful. By 1926, engineers with what was now the Bureau of Reclamation calculated that over 185,000 acres of reclaimed land throughout the West had "become waterlogged, saturated with alkali, or otherwise rendered unproductive" by reclamation.[35]

Whistler and Lewis were more concerned about legal impediments than about natural constraints. Whistler (who eventually worked for Bill Hanley, the major drainage speculator in the county), acknowledged that "the water rights for this area of wild hay pasture and tule land constitute one of the greatest difficulties in the way of developing an irrigation project from the Silvies River."[36] Since the Silvies River water might forever be tied up in legal battles, Whistler and Lewis urged a complicated plan for using "wasted" Blitzen River water to irrigate the basin. They noted that even if water from the Blitzen River were used to irrigate the Blitzen Valley, some water would remain in the river, so

there will probably be an average annual waste of 75,000 acre feet into Malheur Lake. Storage and pumping would make this supply available for use in Harney

Valley. In connection with such a development it would probably be desirable to reclaim an area of 15,000 acres of tule land in the western end of Malheur Lake.[37]

All water, in the engineers' view, needed to be used—not a drop should remain to be wasted on the lake, the refuge, or the birds.

To store one hundred thousand acre feet of water for irrigation of the Silvies River valley, Whistler and Lewis recommended that Malheur Lake be subdivided by dikes, which would also allow drainage and crop-planting on other portions of the lake. They acknowledged that evaporation from such an irrigation storage scheme would be a serious issue: "The evaporation loss from this kind of storage will be so great that it will not be feasible to carry water over to any great extent from a year of high run-off to one of low run-off."[38] Yet that caution did not seem to affect their dreams.

Even grander schemes were afoot. Whistler and Lewis suggested that the federal government dig a canal to drain the lake into the Malheur River, which would return the area to its pre–Great Basin condition of many thousands of years earlier.[39] The engineers suggested that the canal would end annual variations in water levels, allowing managers to fully regulate the water and creating an orderly system that would not threaten crops with annual flooding. Although the canal idea never progressed in 1916, similar schemes to end the basin's flooding resurfaced in the 1980s, driven by the same visions of regulating nature's variations away.

In June 1921, with investors stimulated by Whistler and Lewis's report, the Blitzen River Reclamation District was organized for the irrigation and drainage of fifty thousand acres in the Blitzen River valley. The district proposed construction of a dam for a reservoir of ninety-five thousand acre feet (for irrigation). They organized to reclaim lands from the mouth of the Blitzen thirty-two miles south to the canyon. The consulting engineer, A. J. Wiley, proposed buying all existing drainage and irrigation canals and works in the entire Blitzen Valley, along with all water rights held by Eastern Oregon Livestock Company, which had agreed to sell to the district its irrigation and drainage system in the valley. Wiley proposed completion of the drainage systems and improvement of channels by "removing drift and obstructions in present channel and confining the flow to the regular channel by means of banks built of material excavated or dredged from the river channel."[40]

These efforts stirred concern among William Finley and other refuge protectors. Many preservationists feared that Malheur Refuge, like many other refuges across the West that stood in the way of drainage, would soon be lost. The 1917 report of the Biological Survey noted that "at the Malheur and Klamath reservations, Oregon, deplorable conditions exist on account of uncertainty concerning the status of certain lands embraced within these reservations" and that the refuges should not "be sacrificed for the temporary advantage of a few interested persons."[41] Ira Gabrielson bitterly remarked that Malheur Refuge was at the

> mercy of interests powerful enough to destroy it, while those who had striven for an adequate refuge system stood helplessly by. It is a bitter experience to see the ruin of a biological wonderland, to see it become a desert. It brings a mixed feeling of helpless rage and heartsickness, which must be experienced to be understood. Yet those who knew the wonder of Malheur saw it steadily deteriorate. An aquatic paradise became, successively, a shrinking, withering lake, a stinking mud hole, and finally a barren waste. A similar tragic sequence of events occurred in other places, always in the name of progress. But the destruction of Malheur and Lower Klamath, both sacrificed to the greed of man, seemed especially perfidious because they had supposedly been permanently reserved for the benefit of marsh loving wildlife.[42]

Finley, like his friend Gabrielson, was embittered by the attempt to drain and destroy his beloved marshes at Malheur. He wrote to a friend that "these birds are protected by state and federal laws, and we are in treaty with Canada to protect our water fowl. We arrest a man for killing one bird out of season, but we allow promoters to drain ponds, lakes and swamps under the guise of making agricultural land, and in many cases nothing is gained."[43] In 1916 he orchestrated a series of public writings and campaigns, attempting to draw national attention to Oregon's effort by the state to claim Malheur for drainage.

To win support, Finley combined three strategies: he utilized the emotional appeal of baby animals, he emphasized the economic value of wildlife, and he made an unusual case for the political progressiveness of conservation. His emotional arguments relied on visual images of the plight of young animals. Finley teamed with his friend, the photographer H. T. Bohlman, flooding the media with photographs of adorable chicks, stressing that they would die a slow, miserable death under the drainage schemes. His emotional arguments stressed that Malheur Refuge had unique, immeasurable value for wildlife: "T. Gilbert Pearson, Secretary of the National Association of Audubon Societies, who inspected the Malheur Lake Federal Bird Reservation in August last, has made this statement that it is the most important breeding place for wild fowl of which he has any knowledge."[44] Finley's use of this strategy showed how much he believed, or at least hoped, that national opinion had come to favor wildlife, especially adorable wildlife.

Finley, following other conservationists of the day, argued for conservation's value for human economic progress. His argument centered on the reasoning that bird conservation, by supporting spending on recreation by hunters, could bring in more economic value to the Malheur Lake Basin than could agriculture. Finley quoted the president of the National Association of Conservation Commissioners, writing that drainage of Malheur would prove "unwise economically, because it substitutes for a certainty of valuable birds, the uncertainty of agricultural products on alkaline soil, by nature unadapted for agricultural purposes."[45]

Finley's third strategy was more unusual and more difficult. Knowing that reclamation efforts had been driven by dreams of political progressiveness and social justice, he tried to appropriate these dreams for preservation. He argued that reclamation only *appeared* to favor the poor, whereas in actuality it was a scheme by the rich to get richer, through "exploitation of a public asset for private gain."[46] He wrote that "promoters settled debts on unsuspecting farmers, for they encourage eastern people to settle on alkali land where it is impossible to make a living."[47] Finley believed that developers and speculators, in their efforts to transform marshes into suburbs, were in league with railroads, chambers of commerce, and the rich. He wrote in one article,

> civic organizations, chambers of commerce, and even the railroads fell in with the idea of inducing settlers . . . to come out and locate on remaining public lands through the dry sagebrush country, and especially sub-marginal areas. Common sense would have told even a casual observer these were not fit to support families. So this promotion fever swept through the western country, eating away the feeding and nesting places of the migratory flocks. Water birds could not live without homes. This false land promotion was an epidemic that would take everything in its way, not merely kill a few birds for their plumes. And woe to the next generation.[48]

Other advocates for the refuge picked up on Finley's progressive rhetoric, hoping to persuade locals that reclamation would not help the poor. George Willet, an inspector for the U.S. Biological Survey, spent four months touring the refuge lands in 1918, first trying to understand and then to influence local opinion about the federal government's conservation work. Willet attacked the drainage scheme by claiming that it was "designed solely for the benefit of certain real estate speculators . . . with a desire to profit." Not only was the project designed to profit speculators, but those profits would not even stay within the region, for as Willet claimed, "Chicago capitalists are behind the movement." The only way, Willet believed, to restrain the "exploitation of this public resource by private interests" was to have the federal government gain control of the land.[49]

Like many progressives, Willet believed that the interests of working people and conservation were aligned. Willet, unlike Finley, was not a preservationist. Like early reclamationists, he believed that conservation could be aligned with development and that both could prosper through the elimination of wasteful irregularities. Willet argued that the federal government could, through use of proper engineering technology "regulate the water level," protecting the lake for a wildlife refuge, while also saving the farmers around the lake.[50] He wrote,

> I believe that the water level may be controlled in a great measure by a system of dredged channels and headgates. The method of handling this matter would be to keep the hay lands covered with the necessary amount of water until the hay crop is made and the ducks that breed in the hay fields are through nesting. Then the water level would be lowered by opening the headgates so that the hay could be harvested.[51]

Although some engineers who had studied the lake doubted that such regulation was possible, Willet believed that their hesitations arose from political influence, not from valid scientific skepticism. He wrote, "Though I understand a report derogatory to this method of handling the situation has been rendered by an engineer of the Forest Service, I am convinced that this engineer was under the influence of interest[s] who desire[d] to drain the lake at the time the report was made."[52]

Willet's arguments for controlling water reflect a progressive faith in the engineering behind water control. He envisioned a world where the threat to birds came not from the working poor but from the rich, from outside capital, and from an unstable nature. In his report, a long list of threats to Malheur birds included not just Chicago capitalists, local speculators, and greedy ranchers but also erratic snowfall, ravens, coyotes, skunks, and minks (which, he thought, destroyed half of all eggs). The predators, Willet suggested, "might be practically exterminated by systematic trapping or poisoning"; the speculators presented a more difficult but not insurmountable challenge.[53] To save Malheur, Willet argued, conservationists needed somehow to have the State of Oregon cede the lake to the federal government, which had the engineering prowess to manage it for the benefit of all.

Although Finley lacked Willet's faith in engineering, he followed his advice, persuading Governor James Withycomb to recommend to the Oregon legislature in 1919 that Malheur Lake be ceded by the state to the United States for a permanent bird refuge. Finley hoped that this bill would "settle the Malheur question. This bill provided that the State of Oregon would cede any rights that were claimed on the bed of Malheur Lake to the Federal Government."[54] Attacks on the bill centered on William Hanley's slogan "The babies versus the birds!" and few observers believed that birds could hope to win out over babies."[55]

William Hanley—rancher, speculator, drainage promoter, and self-styled conservationist—worked hard to defeat the bird bill. In his papers at the Oregon

Historical Society are a collection of quotes that he gathered from experts, as part of a press release to the media titled "Opinions of People Who Know the Facts about the Roosevelt Bird Refuge." These quotes reveal some of the reasons that many Oregonians opposed the bird bill and the refuge. First, many people still believed that the basin could be turned into productive farmland. They based their beliefs not on soil samples but on the evidence of their senses during good years (unburdened by their memories of poor years). For example, one Frank Davey of Salem wrote,

> There is a little bunch of grain at the state fair that every one in Oregon should see. This grain was grown by Ben Ausmus, a squatter on the Malheur lake in Harney County, which Finley and the Audubon society wish to have the state surrender to the federal government. The claim is made by Finley that this tract of 30,000 acres of land is a barren alkali waste—non productive, and of no value, except as a nesting place for birds. The grain that Mr. Ausmus brought is an eloquent contradiction of that claim. Mr. Finley has aroused in the children of the state a fear of bird destruction that is groundless.[56]

Or, as the *Portland Spectator* expressed it, in stark and biblical terms,

> The Spectator is heartily in favor of bird refuges, but is much more in favor of bringing into cultivation the small fraction of Central Oregon lands that can be cultivated, even if the entire Malheur reservation should thereby be destroyed. Let us not by our votes doom Eastern Oregon to remain a howling wilderness. Vote No on the bird bill and give the struggling farmers of Harney Valley a chance to exist as well as the birds.[57]

State's rights were a key issue in the fight, for many Oregonians were reluctant to grant more power to the federal government. A firm of irrigation engineers in Portland argued that

> the proposed measure if it passes will make future development of the Harney Valley impossible. Control over the waters which is now in the hands of the state, will pass to the federal government. There is a question as to the constitutionality of such a measure, but in any case the least that can happen is to throw the water rights affected into litigation, and adversely affect the irrigation districts which are now organized and planning for early construction.[58]

These engineers were correct in that the intent of the bill was to harm irrigation districts under construction. Ironically, perhaps, for a man who played successfully on local fears that the state might lose control to the federal government, Hanley called on national engineering organizations to show Oregonians that real experts in big cities opposed the bill. He made certain that the issue attracted national attention among irrigation engineers, and he persuaded the American Association of Engineers to throw their profession's weight behind the opposition.[59] Hanley's efforts were successful. Voters in the state general election of 1920 soundly defeated the bill, much to the dismay of Finley and the Biological Survey.

The governor attempted to negotiate a compromise to protect the refuge while also ensuring development. He appointed a committee of state and Bureau of Biological Survey agents to recommend "suitable action to be taken by the United States and the State of Oregon which would result in the settlement of the Lake Malheur controversy and permit the irrigation development of that region to proceed."[60] Even though the Biological Survey was part of the committee, the terms that the committee recommended were, as the governor intended, biased toward the state and its development plans. The committee recommended that development of the Silvies River reservoir and irrigation plans on the north side of the lake proceed, and that development on the south shore of the lake proceed as well. All prior holders of water rights should continue to receive water. One portion of Malheur Lake within the meander line should be ceded by the federal government to Oregon, and a smaller portion by the state to the federal government. The committee recommended that the state be allowed to erect a dike on the property line and then be entitled to drain all water to a level two feet below the meander line on their side of the lake—essentially, taking so much water that the entire lake bed would be dry in most years.[61]

Finley was horrified by this attempt at compromise, and even more horrified that the supposed protector of the refuge—the Biological Survey—had been part of it. Several years earlier, inspector George Willet had written in anger to the survey, "Your bureau made no attempt to support me in actions taken for the benefit of the birds. . . . Had your bureau seen fit to place me in possession of the true facts at the time I was sent to this reservation, many, if not all, of the difficulties that arose might have been entirely avoided."[62] Willet's frustration echoed what Finley bitterly argued: that the Biological Survey was, instead of protecting the refuges, dragging its feet, refusing to support its agents, and playing games with the refuge for its own political ends.

Many locals surprised the state by coming out forcefully against the compromise, ultimately ensuring its demise. The compromise committee had made a serious mistake: they left the squatters along the lake entirely out of the decision-making process. As Inspector Willet had argued a few years earlier, the squatters proved resistant to state efforts to develop the drainage project. The riparian owners along the lake formed an owner's association, with homesteader Fred Otley as secretary, and sent a statement to the federal and state governments declaring that they opposed any compromise and demanding "that the case be settled by law."[63]

When the riparian owners were left out of negotiations, Otley wrote directly to the chief of the Biological Survey, E. W. Nelson,

> In the Malheur Lake Controversy (U.S. *vs.* State of Oregon) the riparian owner seems to be entirely ignored. . . . Another meeting in the interest of compromise has just been pulled off here at Burns by U.S. and State Commissioners (appointed) at which *we* were not expected. I wrote out the enclosed plea (which is endorsed by the assoc.) and did not have a chance to make use of same, so am forwarding the plea to you with the hope that it may be of benefit to the biological survey and the riparian owners of Malheur Lake whose interest in accord with your expressed view are identical. Please let me hear from you in regards to this matter so we may govern ourselves accordingly—At present as we have always been—we are with you and the bird refuge.[64]

Otley felt that the riparian owners had the same interests as the refuge, revealing how complicated shifting political alliances in the basin had become. In his eyes, both the refuge and the homesteaders could benefit by allying themselves against the rich, on the side of the poor and of nature. Otley attacked the compromise on the grounds that it was "not for the benefit of the Oregon public" but rather would benefit only "the dry lands of a millionaire land and cattle corporation to the detriment of the legal owners of Malheur Lake and at the expense of Oregon's long suffering tax payers."[65]

The riparian owners' association's statement of its demands began by retelling history, for, in conflicts over water rights, history could mean a great deal. Otley described the time before homesteaders came as one of death and stagnation that had been created by nature but worsened by Peter French and other ranchers:

> About the year 1886 said lake with insufficient outlet was dead water, more or less saline, and all the lands above and below the present meander line of lake were waterlogged for want of proper drainage and covered with an immense growth of tules. . . . [Ranchers] created the *dead water* . . . and Malheur Lake was a worthless body of water unfit for any purpose, agriculture, navigation, or a Federal bird refuge.[66]

Whereas the ranchers fostered deadness, the brave "pioneer settlers" fostered life when they "cut barbed wire fences with an axe, moved in, and took possession of the Lake lands. . . . These pioneers cleared the land by burning tules." According to Otley, this tule burning was what opened the lakes up, causing the waters to flow from Malheur Lake into Harney Lake, creating new life through the action of flowing water. He wrote that "the lake water obtained free access to Harney Lake, and through *drainage* dead water became living flowing water, and grass eventually took the place where tules formerly grew." Tule burning was a natural form of reclamation, he claimed, and the effect was nearly biblical: "Malheur Lake is redeemed by *Flowing Water!*" Pioneers had created both a better nature and a better culture, and the only ones who objected were cattle barons and wealthy speculators, who now tried to claim that redeemed land as their own.[67]

The riparian owners objected passionately to the proposed dike because, like the removal of overflow waters from the lake, it would stop the flushing action of the lake and lead to stagnant conditions once again:

> There would be no drainage—it would be dead water above the dike instead of natural flowing water and tulies [sic] would grow instead of grass—as they did before Malheur Lake was reclaimed. The alkali sediment . . . would be held above the dike and the water would become saline, as is now the case in Harney Lake for want of an outlet. Fish could not live therein. It would damage the Federal bird refuge, and would also impregnate the lake with more alkali to the owners detriment and as afore mentioned not only lake lands but the whole of the Lawen district would become water-logged for want of drainage.[68]

Their central arguments rested on the difference between "live" (flowing) and "dead" (still, stagnant, marsh) water. Because something in the 1880s—a cowboy's boot, a heavy rainfall year, perhaps even settlers' burning of tules, as Otley claimed—had allowed Malheur water to flow into Harney Lake, stagnant nature had changed into a redeemed nature that was better for settlers, for grass, and for birds.

What happened at Malheur illustrates the ways that conservation often depended on local people's efforts and their sense of justice, as Richard Judd argues in *Common Lands, Common People.* Judd's work demonstrates the ways in which ordinary people, through their efforts to define the morality of resource use, helped shape America's conservation legacy. Debates over fair access to resources were not purely elite exercises. Locals felt passionately about such issues, and their involvement was critical for shaping the emerging refuge system. At Malheur, for example, the locals with the least power—the riparian owners who had been squatters on French's contested meander lands—joined with preservationists to save Malheur from the state compromise.

This local opposition led to the collapse of the compromise, and the case went to court to determine exactly who owned the lake bed. For the next fifteen years, courts argued the question. Finley wrote with great anger of this time,

> Next the birds were hauled into court for defective title to their own roofs, and got into an endless turmoil of jealousies and legal and/ors', which was settled only yesterday in favor of the government. This was the contest between the State of Oregon and the United States as to the rightful ownership of the bed of Malheur Lake. The state, although it was in no position to patrol the lake or protect the birds, has held out for years. It had cost the people a goodly sum of money, this dawdling in the courts.[69]

As the attorney for Oregon, L. L. Liljiqvist, wrote in a statement to the state Land Board in 1934,

> This case culminating in the present lawsuit has lasted from 1919 to the present time. . . . In the meantime we have had a succession of governors, state treasurers and secretaries of state, and as a result with the succeeding administrations there has been some confusion and uncertainty as to what this case is all about.[70]

But while lawyers were arguing cases, men and women and drought and dust and fences and birds and storms continued to create increasingly chaotic conflicts outside the formalized battles within courtrooms. After drought and irrigation withdrawals dried up the lake bed in the early 1930s, Finley wrote that

> when the great lake gave its last gasp and lay dry and powdery from one shore to the other, the squatters moved out and took possession. It looked as if they had been watching for the event, for they hustled to get ahead of one another for favored spots, each section marked by a hasty plow furrow. Then tents began to stick up over the lakebed, and soon big tractors were plowing and harrowing the fine, black silt, getting ready to plant crops. They took a long chance, for heavy wind storms blew out seed every year in these wide open spaces.[71]

The trespassers planted eight thousand acres of grain, which thrived until grasshoppers invaded. The relief commissioners sent out workers with carloads of bran and barrels of poison, and they managed to prevent the grasshoppers from completely devastating the crop. Next someone, presumably a rancher, destroyed a mile of the fence that squatters had built to protect their grain crop. When a squatter discovered three thousand cattle within a mile of the "prospective feast . . . the alarm sounded. Men hurriedly drove off the cattle, repaired the fence, and placed armed guards along the full length of the barrier."[72]

These conflicts among ranchers, farmers, and the refuge system over the disputed lake bed threatened to erupt in violence. The Burns businessman and county relief commissioner James Donegan wrote,

> The situation looked serious, but it was pointed out to both sides that they were trespassers, and the question of the title was in the courts, and was claimed by the government as a bird preserve, the State as swamp land and the owners of deeded lands surrounding the lake claimed ownership through the theory of riparian rights. . . . It was finally decided that fence lines covering this 80,000 acre tract would be guarded night and day until the crops could be harvested late in the fall. Each side, stockmen and grain growers, agreed to divide the cost of protection. This was done and the Malheur Lake saved the day for many small farmers and stockmen.[73]

When the local refuge warden, George Benson, tried to prevent the growing and harvest of grain on refuge property, locals reacted with fury. An article in the *Crane American* reported rumors that the federal government was going to "dispossess grain growers of the Malheur Lake bed of their crops and forbid its harvest." In response, Donegan declared such a step "a crime of crimes" and demanded that his congressman complain to the director of the Biological Survey.[74] The warden also fired off a series of letters, memos, and telegraphs to his superiors, trying to goad them into action against the ranchers and growers. For example, in April 1934, Benson wrote to the chief of the Biological Survey, reporting on his experience

> while driving about dry lake bed the other day and checking up on the most deplorable condition I have ever experienced on Malheur Lake Bird reservation. In driving about the once tule islands I notice they are being destroyed by live stock, the old tule growth is being tramped completely away which at one time thousands of birds nested on them. . . . Many other things can be done to help Malheur Lake regain her once famous sanctuary. The season is going fast, and it is time something was being done.[75]

But very little was done, for few in Washington were willing to risk bad publicity in the midst of court battles. The squatters harvested their grain off the lake bed, and irrigation withdrawals continued to divert water from the lake.

The pressures from drought, farming, ranching, and loss of habitat led to tremendous die-offs of birds on the refuge. Finley wrote,

> Mr. George Benson of Voltage, who has been acting as game warden there for many years, wrote me that millions of ducks died there last summer and fall. I asked Mr.

Benson how he estimated and he said that along a stretch of four miles he counted 25 dead ducks every 10 steps. There may not have been millions but there were perhaps hundreds of thousands of ducks that died because of the destruction of feeding, resting and breeding places of water fowl.[76]

Finley blamed the federal government for refusing to fight to protect Malheur. After the acting secretary of the USDA wrote to Finley in 1930 that "the Department is unable to expend any funds in order to assure a proper water level," Finley replied with bitterness,

> There was a small peep of remembrance from the Biological Survey away off at Washington, important seat of administration and protection. They had been playing golf, and otherwise having a good time. For many years they had been too busy to find out much about what was going on at Malheur, or to protect the water rights that legally belonged to them, to see the squatters' fences that soon took the place of plow furrows as boundary lines on the lake bed, or the thousands of cattle grazing in spots where grass grew, the rightful feed for ducks and geese. Of course, they didn't want to crack down on trespassers until the ownership of the lakebed was established or have any fuss with the settlers. This spineless policy in the face of the fact that Malheur Lake Reservation had been established by Executive Order of a President, and that they were in possession and recognized as government property until it was proved otherwise.[77]

KLAMATH REFUGES

As Malheur seemed about to be lost to drainage, Finley watched the nearby Klamath Basin refuges undergo even worse treatment. As in the Malheur Lake Basin, early white settlement in the Klamath Basin had centered on ranching rather than farming. Ranchers had cut marsh hay off riparian meadows, but without irrigation, rainfall was not sufficient for growing most crops.[78] Serious efforts at irrigation started about 1882, and by 1903 approximately thirteen thousand acres in the Klamath Basin were irrigated by private interests.[79] Land speculators urged that the Klamath Basin be considered for irrigation, and in 1903 an engineer from the Reclamation Service estimated that irrigation could water two hundred thousand acres of farmland there.

California and Oregon had acquired Lower Klamath Lake through the Swamp Land Act of 1860, but their efforts to stimulate drainage and reclamation had failed. In 1904 and 1905 California and Oregon ceded the lake back to the federal government, for use by the Reclamation Service, to whom Oregon gave the right to the water of the Klamath River.[80] In February 1905 Congress approved the Klamath Project, and work began.[81] The project was to include some dramatic reengineering of the basin: two lakes (Tule and Lower Klamath) would be dried up so the land under them could be farmed. The government would then construct—behind Clear Lake Dam and Gerber Dam—two new deeper, smaller lakes to hold floodwaters for irrigation. A dam and canal would divert the Lost River—a river that once was in an interior, closed basin—over to the Klamath River, which drains into the Pacific Ocean. Water from Upper Klamath Lake would be diverted into an elaborate irrigation system.[82] The Reclamation Service would fund construction of irrigation works; people (mostly veterans) would buy land irrigated by those works from the federal government in parcels of up to eighty acres, and they would pay for the land and improvements over ten years. The federal government sold the land but not the water rights to project irrigators, who could simply use the water each year for a modest fee.

Meanwhile, conservationists had discovered the basin's avian abundance. During the summer of 1905, just a few months after Congress approved the Klamath Project, William Finley toured the Klamath basin wetlands. He was awed by what he found, including extraordinary concentrations of pelican rookeries and what he called the "greatest feeding and breeding ground for waterfowl on the Pacific Coast." By 1908 he persuaded President Roosevelt to create the Lower Klamath Lake Wildlife Refuge, to preserve nesting grounds for migratory waterfowl. This was to be one of the largest wildlife refuges ever authorized, one of the first on land of any agricultural value, and the first in a watershed being transformed by the Reclamation Service.[83] In 1911 President Taft established the Clear Lake National Refuge, and in 1928 President Coolidge established Tule Lake National Wildlife Refuge. The Biological Survey would manage both, and land therein would not be made available for settlement.

Roosevelt's designation created inherent conflicts. These refuges were to be managed by the weak Biological Survey, but they were ultimately under the control of the Reclamation Service. To the Reclamation Service, wetlands and riparian areas were not extraordinary bird habitat; they were wastelands waiting for reengineering into productive agrarian machines. Although the agency was ordered to manage the wetlands for wildlife habitat, its real goal was to drain them, knowing that that would decimate the refuges. As Finley argued, having Reclamation control the refuges was akin to having the fox guard the chicken house and telling him that by law, he could protect chickens only after he had eaten his fill—since feeding foxes was obviously better than breeding chickens.

Roosevelt had intended no settlement within the refuge, but the Reclamation Service skirted that interpreting the refuge boundaries to encompass only land that was covered by water year around. This meant that lakes and wetlands drained by the Bureau of Reclamation would no longer be refuge land and would be available for purchase.

Before draining the first refuge, Lower Klamath, the Bureau of Reclamation commissioned soil surveys to see if the area would be good farmland. Dr. C. F. Marbut, a government soil scientist with the Department of Agriculture, completed a report that concluded the lake bed would be utterly worthless for agriculture. Nowhere in the world, his report argued, had claylike, colloidal soils such as those under the lake ever been made to produce crops. "We can not cite an example of the successful cultivation of a soil of similar character," admitted Copley Amory, an economist with the Reclamation Service, in response to this discouraging report.[84] Moreover, Marbut wrote, wetlands surrounding the lake would have only a slim chance of agricultural success, since after being drained, the peat would be subject to smoldering fires and subsidence. Possibly, Marbut concluded, the peat soils might be grazed, and perhaps be made to produce enough income to pay for the ditching, drainage, sumps, and irrigation.[85]

The Reclamation Service ignored this discouraging report and decided to spend nearly $300,000 on a scheme to drain the refuge. Conservationists were horrified when in 1915 President Wilson reduced the Lower Klamath Lake National Wildlife Refuge from 80,000 acres to 53,600, freeing up the rest for drainage and sale.[86] In 1917, the Reclamation Service awarded a contract to the California-Oregon Power Company (COPCO, now Pacific Power and Light) authorizing construction of Link River Dam for power generation. That same year, 3,000 acres of project land were opened for homesteading. Dikes blocked part of the flow of water to the lake, thus drying up 85,000 acres of rich wetland and riparian habitat. As the wetlands shrank, the Klamath Project opened the land to soldiers returning from World War I.

Finley, in despair, wrote,

Man has a peculiar habit of building something with his hands and, at the same time, kicking it to pieces with his feet. In no direction is this more graphically illustrated than in conservation and use of natural resources. With grandiloquent inconsistency we move to conserve or develop one resource while, at the same time, we are destroying another. . . . No more graphic example of this can be found than the destruction in the name of reclamation of important areas in southern Oregon.

In fury, he added,

Laws have been ineffective because of the schemes of land promoters. . . . Nearly 100 million acres of such drainage has taken place . . . a very large amount proved useless for agriculture. . . . The expenditure of $283,000 proven to be a failure by the Depart of Agriculture, and of no value from an agricultural standpoint . . . relinquished to Klamath Drainage District as a land promotion scheme.[87]

He watched Lower Klamath Lake dry up until just a few pools remained on the refuge. Great flocks of waterfowl were forced to concentrate on those tiny pools, and disease rates shot up, killing millions of ducks in one month alone at Klamath. The peat beds of the vast wetlands begin to burn and collapse, farm efforts failed, and homesteaders went bust one after another. By 1925 nearly everyone involved agreed the attempt to farm Lower Klamath Lake was a dismal failure.

Even after the Bureau of Reclamation drained each lake and wetland for irrigation water, it then put what remained of the refuges up for grazing leases, which further infuriated Finley and other Biological Survey employees. Finley's friend, the ornithologist Ira Gabrielson, described the situation at Klamath in 1920:

> The water table on the lake has been lowered several feet by closing the gates which control the inflow from the Klamath River. This action, made under agreement with the water users' association, has uncovered large areas of alkali flats without thus far benefiting the settlers adjoining the lake or opening up additional land suitable for agriculture. . . . Its future as a refuge is seriously jeopardized. . . . This is an understatement of the wildlife tragedy involved in the loss of one of the two greatest waterfowl refuges then in existence.[88]

Finley's papers at Oregon State University reveal that he wrote nearly fifty letters to various Bureau of Reclamation and Interior Department employees in two months alone, trying to change refuge policy at Klamath, and they wrote increasingly irritated letters back to him, reiterating their positions. In one typical letter, hardly designed to win support from the bureau, Finley wrote,

> Today the records show that the Bureau of Reclamation . . . have killed far more water fowl than all of the market hunters combined because they have destroyed the vast breeding, feeding and resting areas. With the prestige of the government behind it, this Department has played the double role of land shark and land lord. Land is sold outright to farmers at exorbitant prices.

Furthermore, Finley claimed, the Bureau of Reclamation knew that the land would never produce crops, so its intention was merely to make speculators rich.[89] Finally Mr. Hayden, the reclamation superintendent for Klamath, told Finley and the Biological Survey that he might be willing to keep cattle out if the Biological Survey paid him and the ranchers the value of the grazing leases. Finley exploded, writing to newspapers that "Mr. Hayden wants the government to pay the government for conserving waterfowl resources on a federal wildlife refuge." The Biological Survey paid for a fence to protect nesting areas on Clear Lake, but sheep were let in by the Bureau of Reclamation, herded within the fence, thereby trampling the few birds that had started nests. Finley continued his struggle, in his letters trying to attack the Klamath Drainage District for nonpayment of debts and for illegal grazing leases. Try as he might, he could never gain enough evidence to convict the district of fraud, and drainage continued. As Finley watched his beloved Klamath refuges succumb to reclamation, he despaired. He used this despair, however, to refuel his dedication to save Malheur through the purchase of land and water rights—even though privately he wrote to colleagues that he feared it was past saving.

SAVING MALHEUR

The Malheur and Klamath refuges' problems with drainage were not unique. By the late 1920s biologists realized that intensive drainage was destroying critical habitat for avian feeding, breeding, and migration throughout the continent. The refuge system did not offer much help, since it was poorly funded, understaffed, and often subject to drainage. Bills that had been introduced in Congress in 1921 and 1924 to fund refuges with hunting license fees had been defeated. But in 1928 enough national concern had accumulated over waterfowl that when South Dakota senator Peter Norbeck introduced another refuge bill, he finally managed to win approval for it. Norbeck's bill established the Migratory Bird Conservation Commission to acquire wetlands. But funding for wetland acquisition was not available. Federal agencies had contradictory policies as well: whereas the Biological Survey tried to protect wetland breeding areas, policies within the Department of Agriculture and the Army Corps of Engineers promoted drainage.[90]

Motivated by warnings of drastic declines in waterfowl populations, Franklin Roosevelt created the special Committee on Wildlife Restoration to study the problem, appointing Thomas Beck (a journalist), Ding Darling (a cartoonist who had been involved in wildlife conservation in Iowa), and Aldo Leopold to the committee. The men reported back with a condemnation of drainage: "There is incontrovertible evidence of a critical and continuing decline in our wild life resources, especially migratory waterfowl, due to the destruction and neglect of vast natural breeding and nesting areas by drainage, [and] the encroachment of agriculture." The ultimate cause of the problems, the committee argued, was a misguided notion of progress or, in the report's vivid phrasing, "the random efforts of our disordered progress toward an undefined goal."[91]

The committee urged that $25 million be allocated to "restore submarginal lands as wildlife refuges," and Roosevelt promised $1 million to begin the project. Fighting a losing battle, the committee urged that restoration required, first and foremost, planning and coordination so that one government agency did not destroy wildlife to create agricultural surpluses that another agency was trying to halt.[92]

Roosevelt appointed Ding Darling, a close friend of Finley's, to head the Bureau of Biological Survey, and Darling transformed the poorly funded and poorly managed bureau with an infusion of energy, fund-raising skills, and scientists.[93] Most important, Darling helped gain congressional approval for the Migratory Bird Hunting Stamp Act of 1934, a law that financed refuges by authorizing the sale of duck stamps to hunters. Enlisting the aid of local women's groups and sport hunting clubs, Darling planned a string of refuges along the Pacific Flyway, the migratory route for much of the continent's waterfowl. President Roosevelt and Congress stalled, however, diverting duck stamp money to other programs.[94] Darling scrambled for money and finally found it when his ally in the Senate, Peter Norbeck, won $6 million for the refuge program in 1935.

With the hope of gaining federal funding for land purchases, Darling, as new chief of the Biological Survey, began to investigate ways to save Malheur Refuge. Irrigation and drainage projects along the Silvies and Blitzen Rivers allowed very little water to reach the lake, and the biologists feared that winning court cases over title to the lake bed would accomplish nothing if the refuge had no water rights and therefore no water. The Silvies River supplied much of the water in Malheur Lake, but Finley and Darling decided that trying to acquire those water rights would be impossible, for they were divided into many separate holdings and tied up in various court battles. Instead, Finley turned to the Blitzen River, which was still controlled largely by one corporation.

Darling and Finley pushed for federal funds to purchase the Blitzen Valley from the current owners, the Swift Corporation (meatpackers). In May 1934 Darling wrote to Finley that he and Swift had come to a "very amicable understanding regarding the Malheur Lake and the Donner and [sic] Blitzen region." Darling was certain that he could purchase the lands "if we can get the promised funds liberated from the Federal Emergency Relief Corporation."[95] Darling's hopes were soon dashed, for within the week those federal funds vanished, and the Swift interests, tired of waiting for federal action, began selling off sections of their holdings in the Blitzen to other buyers. As Darling wrote to Finley, the Swift interests were "pressed for funds. . . . I am very much distressed that we

can not act at once."[96] Finley replied to Darling ten days later in despair, certain the deal had collapsed,

> Approximately a hundred and sixty thousand acres, the cream of great breeding
> and resting places for water fowl on the Pacific Coast, have been completely
> destroyed. . . . I realize that you feel the same as I do and that you are doing every-
> thing possible, but the thing seems hopeless—at least for the present season.[97]

After decades of frustration—squatters, legal battles, drainage efforts, drought,
and vanished funds—the log jam suddenly broke. The Federal Emergency Relief
Program released funds for the purchase, and on September 25, 1934, Swift agreed
to accept $675,000 from the federal government for sixty-five thousand acres of
the Blitzen River valley, "with all water rights attaching to said lands estimated
at about 150,000 acre feet per annum."[98] Just when Darling found money to buy
the Blitzen, the Supreme Court ruled in favor of Malheur Refuge on the lake
bed title question, finding that the lake was not navigable, so the State of Oregon
had no claim to the lake bed or to the water. The Supreme Court based this decision on the decision in *Marshall* v. *French,* the case that had led to a triumph for
squatters against ranchers while also establishing federal rights to the lake.

In a wonderful irony, the West's grandest cattle empire became its grandest
duck and wetland empire. Whereas local papers were bitter—the *Crane American*
predicted angrily that "loss of these areas from the taxable land would break
the county"[99]—urban papers focused on the romance of the old cattle kingdoms. The *Oregon Daily* summed up the urban feeling: "Where once the wild
yells of savages and the shots of gunfighters resounded, henceforth only the
muted calls of nesting waterfowl will break the silence of the plains, and the 'P'
ranch, scene of the last stand of the old West, will pass into the limbo of peace-
ful pursuits."[100] The *Portland Oregonian* waxed even more nostalgic in its editorial, writing that

> they're going to turn the P ranch into a game refuge and wild life laboratory. . . .
> It seems to us that in the last quarter century, the P ranch has been a sort of focal
> point in a conflict between nature and civilization for supremacy. Now nature has
> won the combat. . . . Now it goes back to nature—in a way. The biological bureau
> of the government has it, probably to keep for all time. Wild birds, of a hundred
> and fifty varieties, will nest in its tules, and game animals will roam its confines in
> safety. . . . But the white-faced steers, and the yipping buckaroos have departed
> these old precincts of Pete French and Bill Hanley forever. Nor may the chugging
> motor cars of wanderers disturb the maternal deliberations of the Canadian honker.
> Nature has won out.[101]

Although urban newswriters could claim that "nature has won out," the reality was far more complex. To save Malheur from the fate that the Klamath marshes had faced, Finley and Darling had turned to politics, money, and law, as well as to the rhetoric of wild nature and romantic cowboys that the newspapers favored. But buying the Blitzen was not enough: to save migratory waterfowl, conservationists soon enlisted the help of a restoration program that borrowed its tools from Finley's old enemy, reclamation.

4 / Managing Ducks

With the goal of restoring waterfowl nesting habitat after the purchase of the old P Ranch, the refuge began extensive engineering projects along the Blitzen Valley for the control of water. Staff bulldozed new ponds for chick-rearing habitat, built dams to hold water, dug ditches for irrigating meadows, and extended a hundred miles of canals along both sides of the valley to supply water reliably to the entire floodplain. Instead of relying on the existing system of wandering channels, which in some years flooded only part of the valley, they hoped to control which meadows were wet, which ponds stayed full of water, and which meadows were allowed to dry out.

Refuge staff built, or tried to build, an empire of nature, a world aimed at increasing waterfowl production. But maintaining this empire led them into continued complications. When carp got into the lake and began making their way up the Blitzen watershed, stirring up sediment and eliminating the sago pondweed that ducks required, staff sprayed rotenone to kill the fish. When beaver returned to the valley and blocked up the irrigation ditches, staff trapped them out, even though the irrigation system was trying to replicate what beaver had created in the first place.[1] When muskrats and badgers began digging holes in the earthen dams, staff trapped them too. When willows began recuperating and spreading along the ditches and canals and channelized banks of the Blitzen, they had to be poisoned as well, since staff felt the shrubs were getting in the way of water management.

Malheur is not the only example of such management; similar puzzles exist throughout the federal refuge system. I started the research for this book with a simple question: Why did managers make decisions that now seem so destructive? I initially hypothesized that these heavy-handed behaviors in riparian management were a product of a particularly American faith in engineering. In *Conservation and the Gospel of Efficiency*, historian Samuel Hays argues that progressive conservation aimed not just to protect resources but to place management in the hands of scientific elites who could engineer the most efficient use of those resources. Progressive managers shared a firm faith that science would allow them to understand all there was worth knowing about the world. Redesigning wild nature as an orderly, efficient machine was at the heart of their efforts. As one government scientist wrote, "The course of nature has come to be investigated in order that it may be redirected along lines contributing to human welfare."[2] Hays shows that conservationists firmly believed that science, conscious purpose, and human reason could engineer a better world.

What I found at Malheur, however, was something much more puzzling. The men who put these actions into place were not managers with a great faith in engineering progress, as I had expected, but rather fierce preservationists who had been the strongest opponents of drainage, ditching, and engineering wetlands. They were people like William Finley, who was fundamentally opposed to an engineering worldview of progress, decrying what Roosevelt's Committee on Wildlife Restoration called "the random efforts of our disordered progress toward an undefined goal."[3] Yet during the Great Depression, preservationists and refuge managers set into motion policies that eventually led to what they most deplored.

To understand their decisions, we need to remember the desperate conditions of migratory bird populations in the first decades of the twentieth century—and the desperate attempts ornithologists and conservationists were making to save those birds. In the early 1930s severe droughts desiccated wetlands along the Pacific Flyway that were not already being ditched and drained. By 1934 the continental waterfowl population dropped to a low of twenty-seven million birds; only 150 egrets and 14 whooping cranes remained. Conservationists began to predict mass extinctions. The bleakness of the losses at the Klamath refuges in particular led Finley and Ira Gabrielson to advocate what were basically engineering solutions for restoration of Malheur—the same techniques that they knew had devastated the marshes in the first place. What had happened to Lower Klamath Refuge, Upper Klamath Lake, Clear Lake, and Tule Lake—and what seemed about to happen to Malheur Lake—filled Finley with fury and despair and convinced him that preservation of habitat alone would ultimately be powerless against land speculators, profit, and engineers bent on creating farmland out of wetland.

In 1934 the Committee on Wildlife Restoration of Aldo Leopold, Ding Darling, and Thomas Beck had concluded that active restoration programs would be necessary to restore waterfowl habitat and that money was the necessary (but not sufficient) tool for successful restoration. The committee called for funds to make possible the construction of dams and dikes, fencing and ditching, excavation and blasting, food planting and land clearing.[4] Even though these tools had destroyed the land through drainage, Leopold, Beck, and Darling believed that they could restore the wrecked landscapes. Ira Gabrielson, in his memoirs about the New Deal years, argued that such programs of active restoration had saved the young refuge system:

> To those conservationists who refused to acquiesce and continued to fight what appeared to be a hopeless and losing battle should go the eternal gratitude of all who love the out-of-doors and the living creatures that are a part of it. Their efforts stirred the public understanding and built up the support which alone made possible the restoration of many of the exploited areas.[5]

To symbolize the effort to restore Malheur, Ira Gabrielson and ornithologist Stanley Jewett drove from the Portland office of the Biological Survey out to the desert, claimed the keys to the diversion ditches from the P Ranch, opened up the diversion dam, and let the irrigation water flood back onto the marshes. Gabrielson said later that this

> was a moment of tremendous satisfaction to see the water flowing into the channel that led to the thirsty lake bed. . . . Malheur again became a great marsh, teeming and throbbing with life as it had been before its destruction. It was a never-to-be-forgotten lesson of the power of man to destroy, and also of the power of man to restore.[6]

Malheur National Wildlife Refuge became the site of the first massive restoration projects in the national refuge system. In radio broadcasts Finley and Jewett announced that "the plan of the government is to make this a super-refuge for the perpetuation of water fowl. On account of location, this will be the most important bird reservation in the West and if the plans are carried out, it will be the largest and best in the United States."[7] At Malheur, with the help of the Civilian Conservation Corps, refuge staff erected dikes and water control structures to impound water and recreate marshes. They built a storage dam for a steady water supply. They created an empire of canals, ditches, channels, water supply dams, and machinery—all to reflood the marshes, to provide the maximum possible duck habitat to keep a few ducks alive.

The refuge system in the drought functioned as a skeleton of habitat, sustaining the minimal needs of the waterfowl. Although some environmentalists might now be critical of the engineering mentality that drove refuge management during those heady years, Finley and Darling felt the situation was so dire as to leave little alternative. In the eighteen months that Darling ruled what soon became the Fish and Wildlife Service, he prevented what the historian Ann Vileisis called the "complete demise of continental waterfowl populations" while setting into place "institutions that would protect wetland habitat in the future."[8] Darling and Finley saved the day for ducks, but they left behind a culture dedicated to heavy-handed solutions.

WINNING OVER LOCAL OPINION

Shortly before the sale of the Blitzen Valley holdings to the refuge system, Stanley Jewett became superintendent of Malheur Refuge. During the year and a half that he held this position before being promoted to regional director, he oversaw the beginnings of the water control program. On October 1, 1934, he opened dams and ditches in the Blitzen Valley, trying to reflood the lake bed. He wrote of this in a letter to Ding Darling:

> The next day the Wm. Dunn dam on the Blitzen River was opened wide thereby releasing some stored water which flowed into the lake bed. The next day the Grain Camp dam on the Blitzen River was opened by Swift & Company's Ranch Manager and A. H. Page, company engineer. This released practically all of the available stored water in the Blitzen River and immediately caused a considerable area within the lake bed to be flooded.[9]

Jewett thought his largest task was to control the flow of water on the refuge. He soon realized, however, that he could not control the water without gaining some control over the neighbors, who proved far more unruly than floodwaters had ever been. Almost immediately after he arrived, Jewett began to struggle with local sabotage of his efforts at water management. Squatters opened his dams, poked holes in his levies, and dug out his traps. In a letter to Darling, he detailed his frustration:

> On the 11th the three of us made a hurried inspection trip over the entire Malheur-Blitzen River area. We found that one of the river dams had been closed and was flooding wild sagebrush land in that area. After opening this dam Walter Sutherland was employed to act as river patrolman and predatory-animal hunter in the Blitzen Valley for the rest of the winter.[10]

Continuing efforts to undermine the refuge revealed deep tensions that still ran within the community. The homesteader David Griffin remembered how much his family hated the refuge during the 1930s, for they believed it to be little more than a tool of the rich to subdue the poor. Griffin recalled that he and other children had swum across Mud Lake to an island where birds nested, and returned laden with duck, goose, and swan eggs. They shot swans and pickled them in brine for winter use, and trapped muskrat, beaver, badger, coyotes, and bobcat.

Although the refuge lands were important sources of resources for home-steaders who were struggling to subsist, federal regulations forbade such sub-sistence gathering (even though sport hunting was allowed on the refuge after 1949). Griffin saw this as clear evidence that the refuge oppressed the poor in favor of the rich. This is not to say that he or other homesteaders near the refuge were blind to the damages resulting from their activities on refuge lands. Griffin explicitly wrote about homesteaders' effects on the landscape: "Soon the land began to show signs of rapid nutrient depletion. Settlement had started at the end of a wet period and after the land was plowed and overgrazed, the soil dried, the wind took it, and it became powdery alkali dust and sand."[11] What he and other homesteaders objected to was the effort of the government to define access to the refuge lands in ways that excluded them.

Recent works by the historians Karl Jacoby and Louis Warren reveal how wide-spread this tension could be. Conservation could lead to an uneasy relationship between the state and locals, particularly when efforts to preserve wildlife turned traditional local uses of animals into criminal acts.[12] As Warren shows in *The Hunter's Game*, the tension between hunters' rights and the government's desire to regulate such hunting helped create resentment against conservation as an elite, urban movement.

At Malheur these tensions intensified after the Supreme Court ruled that the state of Oregon did not have ownership of the lake bed but that homesteaders' claims still needed to be adjudicated in court before the federal government could take possession.[13] The federal government made offers to buy the squatters out, but conflicts over fair prices for refuge inholdings continued for years. Finley claimed that "to be fair and avoid criticism, offers double the actual agricultural value have been made, and most of the area needed has been purchased." He accused unscrupulous lawyers of manipulating the squatters by persuading them to hold out for "exorbitant" prices.[14] In turn, Congressman Walter Pierce accused the federal government of attempting to coerce squatters into accept-ing low offers. Furious, Henry A. Wallace, secretary of agriculture, replied,

This Department is doing all within its power to retain the confidence and coop-eration of the residents of Harney County notwithstanding continuous efforts of others who seem to be opposed to the Government's appropriation and use of this reservation area for the National conservation of migratory waterfowl in fulfillment of the Government's treaty obligations.[15]

Because some government attorneys feared that buying out some squatters would imply a recognition of other squatters' claims to title, the government stopped purchasing land. Homesteaders who had settled on land within the meander line brought suit against the federal government, eventually winning their cases. The government then tried to condemn properties and buy the land, but the prices were too high, and purchases stalled. By 1939 Malheur Lake was still administered by the court receiver, and the protracted legal battles prevented the refuge from, in the new refuge manager John Scharff's words, getting "full control of the Malheur Lake unit."[16] Eventually the Department of the Interior removed the squatters but ironically compensated them enough to set them up as cattle ranchers. Homesteaders, who had for decades portrayed ranchers as the scum of the earth, were perfectly happy to become ranchers themselves when the opportunity was available.[17]

William Finley, Stanley Jewett, and other Oregon biologists realized that a campaign to control water would not save Malheur. Recognizing that a campaign to control local opinion was equally important, they chose two strategies: first, they would address economic concerns, and second, they would attempt to change the language of the conflict, borrowing the rhetoric of the homesteaders to portray the refuge in new terms.

Much of the opposition to the refuge had been, at least on the surface, on economic grounds. Locals had warned that the refuge would take land out of the county tax base and also put an abrupt halt to development. Finley and other advocates saw their task as persuading locals that the refuge could in fact be a boon to the local economy. Finley wrote that "some people have always complained that where an area is set aside as a federal wildlife refuge, it takes land out of use for grazing and cultivation." But this was wrong, he claimed, for refuges actually helped subirrigate the surrounding lands, thus improving surrounding farms. In contrast, he claimed, attempting to farm the lake bed was false economy, for it would lead only to desertification:

> The most vital need in this area is a good supply of water. Without water, the bed of Malheur Lake might be cultivated for one or two seasons. Then it would revert to a desert and be a loss both to men and birds. If a good part of the lake bed can be filled with water and this level maintained, all the land surrounding will be made useful for grazing by sub-irrigation. This will benefit the bona fide owners around the border of the reservation.[18]

Finley tried to win local opinion over to the refuge by promising indirect grazing benefits, but these were not nearly as persuasive as direct grazing leases. The need to win over local opinion and to fund the restoration projects soon resulted in an extraordinary grazing and haying program. Even though Finley loathed the idea of cattle grazing on refuges, and even though he had fought it on Steens Mountain and on the Klamath refuges for years, at this point in the restoration of Malheur, he conceded that allowing cattle grazing might be the only practical way of getting funds and public approval for restoration. He grudgingly admitted that

> by handling this property in a practical way, the government will derive an income which will help with the upkeep. A certain amount of stock can be grazed, and some of the meadows will produce a good supply of hay. The season of grazing and cutting of the grass will be governed so as not to interfere with the breeding birds.[19]

Initial grazing and haying programs on the refuge were designed partly to raise funds for restoration projects, but more important, they were intended to lessen local hostility to the refuge.

Finley's second strategy for winning local opinion was to tap into the powerful dreams that had motivated homesteaders and ranchers. He had come to believe that restoration's chief enemy was not just drainage itself, but the power of the dreams that drainage had made seem possible. Finley and other conservationists felt that they needed to create a compelling vision that could compete. To craft this countervision, they tapped into both the romantic myth of the Wild West and the powerful Progressive dream of equality and social justice. When Finley wrote one of his most impassioned appeals for conservation in 1923, he appealed to economic hopes and to the desire for social justice, writing that "while Roosevelt was a great advocate of drainage and irrigation through the arid sections of the West, he also understood that the bird life of the nation which belongs to all the people, was an economic necessity."[20] His article focused on the human devastation caused by unscrupulous drainage developers, stressing how corporations had manipulated American visions of social justice by tricking the working poor into a life of terrible toil. A friend of Finley, the writer Dallas Lore Sharp, described this life as little more than a slow death:

> I have seen many sorts of desperation, but none like that of the men who attempt to make a home out of three hundred and twenty acres of High Desert sage. For this is so much more than they need. Three feet by six is land enough—and then there were no need of wire for a fence, or of a well for water. Going down to the sea in ships or into mines by a lift, are none too high prices to pay for life; but going out on the desert with a government claim, with the necessary plough, the necessary fence, the necessary years of residence, and other things made neces-

sary by law, to say nothing of those required by nature and marriage, is to pay all too dearly for death, and to make of one's funeral a needlessly desolate thing.[21]

Refuge advocates tried to counter the dreams of drainage and homesteading not just by portraying homesteading in such dour terms but by creating a vision of Malheur as an empire of nature with the grand romance of the lost cattle kingdoms. In a 1935 radio talk, Finley described Malheur as "almost an empire in itself, of 164,000 acres of what is conceded by naturalists, sportsmen, and conservationists as the greatest single wildlife area in the nation." All the "romance of the old West is still much in evidence on one of the nation's greatest wildfowl refuges," he argued, going on to describe how

> sixty years ago a far-sighted young Californian, Pete French by name, in search of good cattle range wandered into what is now known as the Blitzen Valley, a wide flat plain watered by a fine stream, green with wide meadows of luxuriant grasses, interspersed with thickets of willow, and with great areas of swampy ground and shallow ponds.[22]

In borrowing the myth of Peter French and the cattle barons for Malheur, Finley was trying to create an empire of ducks that would have some of the emotional and rhetorical appeal of the cattlemen's empire. Yet the strategy backfired for two reasons. Squatters, who had earlier aligned with the refuge in their struggle against the ranchers and the state of Oregon, were not likely to find this an appealing myth, and indeed they soon abandoned support for the refuge. The second reason that the strategy backfired was more complex: the myth seemed to have had its greatest power not so much among homesteaders and other refuge critics but among the refuge staff themselves, in ways that helped to shape their own decisions about their manipulations of the empire they were creating.

Under the four-decade-long reign of John Scharff as refuge manager, refuge staff became increasingly seduced by their own stories of empire-building. Even into the 1970s, they continued to borrow the romance of the Wild West as a powerful trope for the refuge. For example, the refuge marked the 1972 centennial of the P Ranch's founding by publishing an official history that wrote of French as a hero, claiming that he created "what was considered by many to have been the best managed and best balanced cattle empire ever put together in the West."[23] Visions of Malheur as an empire of nature—linked through history to the empire of Peter French, cattle baron—were powerful visions, and they were effective in creating a mythic sense of place for the refuge. But ultimately they were dangerous, for they fostered a mode of thinking that began to believe in its own self-promotions. Acting more and more like the mythic cattle barons as they attempted to manipulate both water and rhetoric, the refuge managers began to see themselves as a world onto themselves, as they turned Malheur into an empire of ducks.

JOHN SCHARFF

John Scharff, the manager of Malheur for nearly half a century, did what Finley failed to do: he transformed local hatred against the refuge into a grudging and then wholehearted acceptance. He did this, in part, by increasing local economic opportunities on the refuge through the grazing and haying programs. Even more important, however, was that Scharff spoke a language locals could understand: he was a rural man who had grown up on a ranch, a man who believed wholeheartedly in the ability of management to increase yields of ducks and livestock both. Although locals often loved him, many biologists still speak of Scharff with anger. Like French, Finley, and Hanley, he was a controversial figure in the valley who instituted a set of complicated changes in the relationships between land and water.

When Scharff came in as manager, he had hoped to focus attention on Malheur Lake itself as the centerpiece of the refuge. As he recalled, "The first thought was to divert all Blitzen Valley water into the lake area but at about the time more troubles were looming in the future with regard to the actual ownership of the lake bed of Malheur Lake."[24] Continued troubles over the title to the lake bed, however, meant that Scharff soon turned his full attention to the Blitzen Valley. This had been acquired for the rights to water that would refill Malheur Lake, not for the river valley's spectacular riparian areas and meadows. But because the lake bed was still contested territory, refuge staff saw new value in the riparian habitats. The result was that soon, in Scharff's words, the Blitzen Valley Unit comprised "vast acreages of natural meadowlands and willow bottoms, interspersed with small swamp and open water areas mostly as a result of the canal and dike system provided during the early development of this particular refuge."[25] This habitat diversity was partly natural, but it was also partly the result of Scharff's work.

In 1937 refuge staff, along with men from the three ccc camps located on the refuge, built over 150,966 cubic yards of levees and dikes, set 95 miles of barbed wire, cleared out 83,938 cubic yards of channels, laid 34,680 cubic yards of riprap, and set out 35 separate water control structures.[26] This was an extraordinary effort of labor, sweat, and machines. By 1937, just a year after Scharff had began work, he could announce in his fiscal report, "The creation of lagoons, ponds, dikes and canals in all parts of the refuge has indeed proven an incentive for the waterfowl to utilize practically all meadows and formerly dry fields as nesting areas."[27] Interspersed with these descriptions were photographs of marshes and ponds with captions such as "Man-made Water Areas where Waterfowl Romp" (see plate 12). Scharff added underneath the photos that "much better use was obtained from the available water this year owing to the facilities which have been constructed to properly handle the water upon its arrival."[28] He summed up this report with the proud statement, "It is gratifying to say the least, to see the increased numbers of birds using the facilities which have been provided by our early development work."[29]

In 1938 Scharff wrote with satisfaction,

> Many of the old water controls of the former cattle ranch regime were replaced and others added at chosen locations. With the completion of many miles of roads, newly constructed canals, bridges, leeves [sic], dykes, and ponds the Malheur Migratory Waterfowl Refuge has outgrown its potential stage and can now be classed as a full fledged member of the Nation's system of wildlife refuges. Although considerable work must be completed before finis may be written to the development program, the year 1938 has been a bumper one from the standpoint of progress, and unquestionably Malheur Migratory Waterfowl Refuge is now and will continue to be a valuable part of the Nation's resources as well as a source of satisfaction and interest to the surrounding country.[30]

What John Scharff was doing with the Malheur Refuge waterways was not unusual for the era. The 1930s were a decade marked by national enthusiasm for wildlife conservation, and much of that enthusiasm was aimed at projects that actively manipulated habitat. In 1934 the Bureau of Sport Fisheries undertook the first nationwide program of stream surveys and habitat improvements.[31] Throughout the West, the bureau began to restore and improve streams on public lands. The program's major emphasis was on structural engineering solutions, what managers such as Scharff termed "improvements." Often using CCC camp labor, between 1933 and 1937 restorationists throughout the nation built a tremendous number of in-stream habitat structures, such as rock dams to create pools for trout, riprap to stabilize stream banks, and deflectors to force streams to meander. Soon many managers came to assume that *all* water sources needed improvement: structural engineering was not just for damaged streams, but for all streams.

The improvement work at Malheur soon moved beyond simply restoring water to meadows that had been dried out by irrigation, to active control of the river's course. The Blitzen River had flooded the P Ranch meadows and orchards practically every spring during Peter French's regime. The ranch employees had dealt with the annual floods by sandbagging important buildings, sometimes by clearing out a few obstructions in the river near the buildings, and sometimes by simply waiting until the waters dropped.[32] John Scharff was determined, however, to make the river behave, so one of his first acts was to have the CCC men build dikes along the Blitzen to keep the annual floodwaters away from the P Ranch house and orchard.[33] In the 1940s, when it was clear that dikes alone were not going to tame the river, Scharff simply moved the river: "Three large gravel bars were moved from the Blitzen River which changed the river channel before the P Ranch house. . . . The river channel was changed in one place where it threatened to cut out into and through the upper P Ranch meadows."[34] When the river tried to move back into its old channel, Scharff did his best to fix it into place, hardening its banks and trapping it within a ditch, cutting off its connections to the surrounding landscape.

BEAVER

When beaver began flourishing, threatening the refuge water control structures in the early 1940s, Scharff had them trapped out. This may sound straightforward: if a burrowing rodent is making your engineering works collapse, kill the critter to save the engineering. But human relationships with beaver on the refuge were much more complicated than that simple decision suggests. Refuge policy toward beaver shifted almost as often as did the river itself, and the increasing rigidity of official policy toward beaver reflects the hardening of Scharff's ideas about controlling nature.

In the 1930s, rather than killing beaver, refuge staff had actually restored them to the refuge. William Finley reflected the beliefs of many biologists of the era when he argued that beaver had

> great value in conserving water. . . . The beaver, in building dams and developing ponds, increases the supply of pasturage for live stock, creates ponds and streams where there is good fishing and recreation in the mountains. The beaver prevents the errosion [sic] of the soil and the supply of water is stored in the mountains so that it comes down gradually and can be used for irrigation.[35]

Conservationists saw beaver as storing water, not wrecking water storage; as benefiting irrigation, not destroying it. In an article aimed at local ranchers, Finley wrote that "killing beaver is in about the same class as stealing cattle."[36] To persuade ranchers that beaver would increase ranching profits over the long run, he used an argument based in history, retelling the story of the decline of Silver Creek (which ran into Malheur Lake from the north):

Years ago, with a good water supply, this was a valuable area for stock. During the winter of 1911 and 1912, two men trapped and took out about 600 beaver pelts from the headwaters of Silver Creek and its tributaries. With no beavers left to keep up the dams, the ponds began to disappear and the water supply lessened each year. Instead of thousands of tons of pasturage for stock, the amount was reduced to a few hundred tons. As the stream dried up, ranchers had to dig wells and pump water during the summer for their stock. Farm lands on the lower stretches of Silver Creek, lacking water, produced nothing.

The trappers in one season had gained about $4,000, but this was like killing the goose that laid the golden egg.[37]

In other essays, Finley speculated about the ways that the riparian landscapes and meadows he loved might have been shaped by a history of beaver:

No man has yet attempted to look back through the ages, study the topography of America and try to evaluate the industry of the untold millions of beaver that inhabited this country. Who can estimate their services in storing water, stopping erosion of the soil, creating the first meadows that later developed into thousands of fertile valleys? In these valleys, new generations of beaver established ponds and marshes for fish, water fowl and other fur-bearing animals. Through centuries of water conservation, the beaver has aided in maintaining the water table which has prevented the land from reverting to a desert.[38]

Finley, like Aldo Leopold, envisioned a new telling of history, an environmental history that would connect wildlife, people, and landscapes.

Finley was not the only conservationist to see the value of beaver for riparian landscapes. In 1935 Stanley Jewett had written to Ding Darling, "The entire channel of the Blitzen Valley is fairly well stocked with beaver while a few mink are found here and there throughout the valley."[39] And Finley had mentioned that year the presence of "a large number of beaver along the Blitzen River."[40] These records indicate that beaver were not extinct or even rare in the Blitzen Valley during the 1930s, even though the drought and irrigation withdrawals had likely reduced their available habitat.

In 1937 the Oregon Department of Fish and Game conducted beaver surveys throughout eastern Oregon. The goal was to coordinate trapping with transplanting, so that nuisance beaver, rather than being killed, could be moved to a place where their activities could restore riparian areas. That long hot summer, two young employees named Merle Markey and Fritz Cramer traveled across the high desert talking with ranchers, searching for beaver, and live-trapping the animals and then transplanting them onto public lands. What Markey and Cramer found was surprising: after the drought years, many ranchers had begun to appreciate the presence of beaver on their land. Rather than wanting Markey and Cramer to take away beaver, many ranchers wanted them to bring in more.

Some ranchers did resent the way that beaver could flood hay fields and interfere with irrigation. Markey and Cramer reported on one ranch in the Silvies Valley that

> beaver here, more than on any other ranch in the valley, have built dams, causing the water to rise and overflow on the hayfields, thus making them too wet and marshy to harvest hay upon. They would dam up the water causing it to flood portions of the fields, thus destroying the hay crop in some instances and causing difficulty in cutting the first crops before the water got lower.[41]

But, as Markey noted, other ranchers appreciated beaver: "On some of the ranches, however, the farmers do not object to having beaver on their places because the ponds conserve drinking water for the stock and often cause the fields to be sub irrigated." Some ranchers felt that beaver actually helped irrigation: "Mr. Paul Stewart, a rancher near Crane creek . . . wants very much to have beaver planted upon his place. There are already 2 dams on Crane creek flowing through his place but they are insufficient to hold back the waters and thus conserve it for irrigating purposes. Mr. Stewart wants as many beaver as he can get to plant upon his place." Many ranchers believed that beaver had both benefits and costs, so a successful operation needed to find a balance between the two. Cramer wrote of one rancher,

> Mr. Cleveland said that in the spring the beaver cause him some damage by flooding portions of his fields and later on, the swampy characteristics of parts of the fields makes it difficult to work them with a mower. However, in the summer, Mr. Cleveland said that the beaver were a distinct aid to him by creating ponds in Wolfe Creek, which flows through his place, & thus conserve the water for his stock. He summed it up by saying that it was a 50–50 proposition and that he guessed beaver were beneficial to him in the long run.[42]

At one point in their summer fieldwork, Markey and Cramer went to Malheur Refuge and talked with John Scharff, who introduced them to local ranchers and suggested sites on the refuge where beaver could be transplanted.[43] Surprisingly, in these early years, Scharff wanted beaver on the refuge, even though by the next decade, he would be doing his best to exterminate the animals.

Soon after the end of World War II, attitudes toward beaver had changed. In 1942 there had been a massive beaver die-off throughout the Harney Basin, perhaps due to tularemia.[44] Four years later, when beaver began rebounding, they became a hated enemy on the refuge. Scharff's 1946 narrative report stated that twenty-eight beaver had been removed in the spring and that "a large beaver house was dug out near Buena Vista and the dike repaired. This was the fourth time for that particular job in as many years."[45] In a 1947 quarterly report, Scharff reported that absolutely all beaver were being removed because "they interfere with water regulation."[46] That year staff and trappers killed forty beaver in the Blitzen; in 1950, thirty were killed; in 1954, forty; and in 1955, fifty more.[47]

Why had beaver become so hated? They certainly threatened Scharff's control of water—they were crumbling the dikes, shifting the foundation of the empire Scharff was constructing. But in the previous decade, beaver had done all that, yet refuge staff, like ranchers, appreciated the riparian benefits beaver gave, while trying to live with their costs. What had changed? To find the answer, one needs to look beyond the confines of the valley.

The post–World War II hatred of beaver in the valley reflected changes in national attitudes toward wetlands and riparian management. As the wetland historian Ann Vileisis has argued, after the war, the Fish and Wildlife Service, the Bureau of Reclamation, the Army Corps of Engineers, and above all, the USDA's Soil Conservation Service all did their best to drain and ditch American riparian areas and wetlands into machinelike landscapes. For the Fish and Wildlife Service, the goal was better duck habitat; for the other agencies, it was better agricultural land.

Across much of the country, drainage had swept agricultural practice during the first decades of the twentieth century, but the Dust Bowl made many question it. After the war, high agricultural commodity prices allowed many farmers and ranchers to begin draining wetlands again. This time, farmers had the help of the federal government, whose programs encouraged farmers to turn marginal areas into croplands.[48] Soil Conservation Service agents considered drainage a fundamental conservation practice and provided farmers with bulldozers and draglines to dig ditches. Moreover, agents considered drainage a way to bring farmers into their network; by helping them drain lands, agents felt they could develop rapport with locals and establish their advisory role for other soil conservation projects, such as those to reduce erosion. The USDA's Production and Marketing Agency shared costs of on-farm drainage projects, paying farmers 60 percent of the costs of drainage. This, along with price supports for surplus crops, meant that federal subsidies effectively removed much of the risk of investing in turning wetlands into agriculture. Drainage became a patriotic mission, part of the post-war dream of using agriculture to feed a hungry world.[49]

In 1954 the Federal Watershed Protection and Flood Prevention Act (known as PL-566) was passed, creating the Small Watershed Program, which would help

USDA agencies work with state and local governments to reduce large floods by damming streams high in watersheds. Because years of erosion had left many streams clogged with sediment, the program was also intended to channelize waterways to carry away floodwaters faster and more efficiently.[50] Since the channelized streams would carry drain waters as well, drainage projects could also fall under this program. By 1955, 103 million acres of land had been organized into drainage systems and $900 million had been spent on ditches, outlets, levees, and pumps. Soon more miles of public drainage ditch than highway covered the country. During four years in the 1950s, drainage funded by the USDA converted 256,000 acres of waterfowl habitat into farms. The technology of the bulldozer accelerated channelization through the 1960s, for the machine made the work quick and cheap. The Soil Conservation Service undertook huge projects of what they termed "stream improvement"—straightening and deepening water courses, removing riparian vegetation, dredging sediments, and thoroughly altering hydrology.[51]

Within the Malheur Lake Basin, projects funded by the PL-566 Small Watershed Program were popular, even in a locale that professed to hate government projects. In 1967 a federal report on the basin recommended nine PL-566 projects for Harney Basin. The report reflected its era, seeing potential improvements and developments behind every sagebrush. The report did recognize that erosion had become a serious problem on streams and rivers altered by agriculture, scolding that "considerable land is lost through streambank erosion." Three times the report insisted that the Silvies River channel badly needed work because "the channel had inadequate capacity, flat gradient, and, in some cases, an almost undefined course." The solution lay in "enlarging, aligning, and clearing these channel[s]."[52] In the planners' view, one reversed erosion not by holding the water on the land longer through revegetation programs but by accelerating channelization (what current stream ecologists would argue is the *cause* of, not the solution for, erosion).

A few wildlife biologists did object to these programs. In testimony to the House Oversight Committee, Nathaniel P. Reed, assistant secretary of the interior for fish, wildlife, and parks, argued that channelization reduced local populations of fish, vegetation, and ducks by 80 to 99 percent: "Stream channel alteration under the banner of 'improvement' is undoubtedly one of the most destructive water management practices. . . . [It is] the aquatic version of the dustbowl disaster. " Bob Scheer, president of the North Dakota Wildlife Federation, argued that "Congress really passed this small watershed law to stop floods where they begin, to hold the water on the land. But instead we seem to be using the law to get the water off faster."[53] Yet these objections mattered little to people eager to rationalize the landscape and increase agricultural production.

In such a national climate, producing ducks became one more agricultural output. When beaver interfered with the drainage, channelization, and irrigation systems needed to maximize that output, they were doing more than threatening a few ditches. They were threatening progress itself. And so, in John Scharff's eyes, beaver had to go.

PHREATOPHYTES

A similar shift in attitude toward riparian plants—from valuing to demonizing them—occurred during the same period. In 1922 Stanley Jewett detailed the numerous species of birds that used willows for nesting, feeding, and cover. Three times in his report, he urged that the refuge staff undertake an extensive program of willow planting.[54] In 1937 John Scharff had the ccc men plant forty-one thousand native willows along the Blitzen. His goal was to stabilize water control structures by "protecting against erosion of roads and dikes," not to provide habitat, as Jewett had advised. Nevertheless, in 1937 Scharff could imagine that woody riparian vegetation might have value. His willow plantings expanded for several years, along with his conception of their utility. In 1938 staff planted 11,764 willows and other trees, and 2,050 more in 1939, with 60 percent mortality.[55] By 1938 he believed that his restored willows could "serve as riprap against wave action, to furnish cover for birds nesting, and also to improve the general appearance of the landscape."[56]

Four years later Scharff had done a complete turn-around, declaring war on willows. In 1943 he wrote that "the Blitzen River channel below the P-Ranch buildings was cleaned of drift and the brush and willows removed from the banks." Along several miles of dike and roadsides, staff replaced willows with the invasive, non-native crested wheat grass "to provide an improved view." In his next quarterly report, Scharff noted that "the rapid willow growth along the channels, canals, and ditches is in need of thinning." In 1949 he began spraying willows with the new herbicides made available after the war. One picture from the report (plate 17) shows a man with no protective gear spraying 2–4D over the water to kill willows along Buena Vista Dike. The original caption happily announces that, along the six miles of riparian area that were sprayed, "about 85 percent of willows ceased growth and lost their leaves."[57]

In the 1950s, herbicide programs mushroomed, becoming an extraordinary focus of effort and pride in the refuge's quarterly reports. In 1955 aerial spraying began with planes released from war service. The September–December 1955 report announced that

plant control work at Malheur Refuge took a new turn during 1955 with airplane spraying replacing outmoded ground methods. Costs per acre were cut from $13.86

in 1954 to $3.42 in 1955. In addition, the weed control activity was completed in two days on 200 acres compared to 25 workdays required in 1954 for 61 acres.[58]

That year, with the help of new technology, "weed control" expanded to include many species of native vegetation, both riparian and upland. The quarterly report noted that

on June 21, 35 acres of desert shrubs east of headquarters were sprayed. The dominant species was big greasewood *(Sarcobatus vermiculatus)* which made up 75 percent of the stand. Big sage *(Artemesia tridentata)* made up 20 percent of the stand, and rabbitbrush *(Chrysothamnus viscidiflorus)* 5 percent of the stand. For this work, Dow Weed Killer was used (formula 40) consisting of ester of 2,4–d plus amine salt. The application rate and degree of dilution was the same as used for willows. Cost per acre was $3.24.[59]

In 1956 staff sprayed 230 acres to rid them of "noxious weeds and willows"; the next year, "brush" was removed with herbicide on 130 acres, and the non-native species crested wheat grass was planted on another eighty acres.[60]

By the 1950s and 1960s weed control became a major objective of refuge administration. Willows were cut, mowed down, and sprayed with 2,4–D for many reasons: to remove predator habitat, increase mowing efficiency in hay meadows, increase the number of acres that could be put into full cattle and duck production, and make it easier for tourists to see the wildlife. One of the most important reasons was to decrease competition for water. Woody riparian plants are phreatophytes, meaning that they extend their roots into the water table and consume a great deal of water. As a 1967 federal report on the Malheur Lake Basin argued, "Many people believe that the high consumption of limited water supplies by phreatophytes is one of the most serious problems in the West."[61]

As agriculture became increasingly industrialized, researchers started looking on riparian vegetation with a new eye. Willows seemed to be draining the creeks, stealing good water that cows and alfalfa could be using. Or, as the 1955 *Yearbook of Agriculture* put it, "Men who have studied the problem throughout the West realize that a large part of the water consumed by phreatophytes could be put to beneficial use by replacing the phreatophytes with crops, grass, or other beneficial vegetation."[62] Phreatophyte removal accelerated with the introduction of new herbicides—the 1955 *Yearbook of Agriculture* recommended that for complete control, one must repeat six sprayings of 2,4–D and 2,4,5–T, which later became notorious as Agent Orange.[63] Water experts of the mid-1950s came to believe that they could create more water and control floods through such phreatophyte eradication programs.

The plan backfired. Without vegetation to naturally impede floodwaters, the resulting higher, faster flows incised channels. The very plants that farmers and ranchers thought drank too much of their water actually had helped maintain a high water table. Riparian hardwoods are thirsty plants, which was often why people cut them down, thinking they were stealing water from livestock and more useful trees. But using water does not always reduce the supply for everything else. Instead, riparian vegetation can allow streams to continue flowing longer. Although they steal water, such plants increase the available supply to other plants. Riparian plants make the boundaries between water and land more complex, slowing water flow and keeping dirt from flooding the streams. Their leaves provide shade, reducing water temperature. Their branches and dead wood falls into the water, trapping debris and forming dams. These debris dams increase turbulence, helping scour deep pools where fish can spawn. Refuge staff used to think all of this was bad—that the point of a stream was to move water from point A to point B as efficiently as possible. But the more people tried to simplify streams by channeling, piping, and cleaning them up, the more the waters dwindled away. Riparian zones made the boundaries between water and land more complex along the Blitzen, and John Scharff, like many other managers, believed that these complexities interfered with his efficient administration of nature.

GRAZING

When the refuge system was first established, no clear national policy on grazing or haying within the refuges existed. That confusion persisted for decades. In 1915 refuge protector Frank Triska wrote to E. S. Cattron, district inspector for the Biological Survey, asking what to do about requests for haying. Cattron replied that Triska should use his best judgment, allowing careful ranchers to cut hay: "About cutting hay on the reserve—use your own judgment in this, giving the ranchers the best of it when you can—only see to it that no birds or nests are destroyed. . . . Until we are advised otherwise let them make hay and pasture stock on the reserve under your supervision."[64]

A year later the new inspector, George G. Cantwell, wrote to Triska that he had "taken up the question" of haying and grazing with his bosses. Cantwell hoped the Biological Survey would allow hay to be cut, "for it is no more than right to the ranchers."[65] A few days later, T. S. Talmer (who signed his letter, "In Charge of Game Preservation") replied to Cantwell, "We have no authority to lease any privileges on bird reservations." He did concede that

> it is possible that arrangements can be made whereby some of the settlers may cut hay if in return they will cooperate with the Department in patrolling the reser-

vation, [but] . . . hay cannot be cut until after the young birds are hatched as the law prohibits disturbing as well as killing or capturing birds on the reservation.[66]

Talmer was clearly dubious about allowing any haying or grazing. Leases were forbidden, he insisted, but informal bargains might be permissible—a little haying in return for patrol work. The intent was certainly *not* to let this become a major program, because in his opinion birds came first.

In 1918 refuge inspector George Willet, recognizing how critical refuge lands were to locals, wrote that

> the swamp land bordering the open water of Malheur Lake is the most important wild hay producing section in Harney Valley and, as such, is of immense value to stock producers of the section. Particularly is this true during a year like the present one, the driest in the history of the region, and the lives of thousands of cattle will be preserved during the coming winter by hay cut on the reservation.[67]

Concerned about winning over local support for the refuge, Willet urged the administration to allow locals to continue grazing and cutting hay on refuge lands, even though no clear policy authorizing such action existed.

By 1920, however, perhaps in response to Willet's urging, refuge grazing policy had changed to something much closer to that of today. The director of the Biological Survey wrote to Malheur warden George Benson,

> It is the policy of this Bureau to allow the public to use the lands under its control to the fullest extent possible consistent with the best interests of the reservation as such. If you feel it will do no harm, this Bureau has no objection to haying operations unless for any reason they prove detrimental to the welfare of the reservation, in which case haying must be discontinued.[68]

Malheur grazing policy was anything but stable during these turbulent years. In 1929 the Biological Survey advised Benson that long-term grazing leases were not acceptable and that only on rare occasions could annual grazing permits be granted:

> In response to your letter of October 30 suggesting that certain lands at the west end of Harney Lake be granted, under permit, to Mr. A. W. Hurlburt for his use in stock raising, you are advised that no authority exists for us to lease reservation lands. However, when it appears that the use of such lands will not interfere with the purposes for which the reservation was established, we occasionally issue permits authorizing the permittee to utilize the land for only a year at a time.[69]

Grazing and haying were *not* to be central uses, and permits were not to be of long duration.

Three months later, in February 1930, policy shifted once again. Benson was ordered by the U.S. game conservation officer from the chief's office in Washington not to allow grazing in the refuge, for cattle seemed to be damaging eggs, destroying cover, and eating food needed for birds:

> Consideration has been given to the subject of domestic live stock grazing Cole Island and other portions of the Malheur Reservation where they consume and trample down vegetation that is important as cover for the bird life, particularly nesting waterfowl. The conclusion has been reached that the Bureau will hereafter endeavor to prevent stock men from ranging their stock on the Reservation. You are therefore authorized to warn stock owners that hereafter they must not permit their stock to range on the Reservation.[70]

Benson replied that although prohibiting grazing might be a fine idea in theory, in practice it would prove difficult, for ranchers faced a desperate shortage of water:

> I believe this is a very important move to keep live stock out of the nesting area, especially in the nesting season. . . . But I realize under the present condition it is going to be a rather difficult proposition to handle. . . . It is a desperate condition where there is a shortage of water for live stock, and the proposition will need carefully [sic] consideration.[71]

A few months later, notwithstanding Benson's concern that it would prove impossible to keep cattle off the refuge during "desperate" times, federal officials reiterated the policy forbidding grazing at Malheur. Stanley Jewett and U.S. game warden Chester A. Leichhardt wrote that the "condition of overgrazing" had destroyed "considerable areas of vegetation which are necessary for suitable nesting cover for waterfowl."[72]

Although damage from overgrazing was obvious during drought years, some advocates for the refuge in the 1930s felt that cattle grazing could benefit waterfowl by keeping "down excessive growth of vegetation."[73] Ranchers soon echoed these arguments, urging the refuge to allow grazing because cattle might improve duck habitat. In the early 1930s few ornithologists agreed. Ending grazing in refuges became a major push for William Finley and other conservationists in the 1930s, particularly after they had witnessed overgrazing effects at the Klamath refuges. In 1934 Finley wrote to Harold Ickes, secretary of the interior, complaining that grazing allowed by the Reclamation Service has certainly raised havoc with water fowl in certain areas of southern Oregon and northern California. Under permits issued by the Reclamation Service, stock has tramped out the nests of game birds and killed off the young birds nesting in colonies on the ground. . . . Ducks and geese live on grass the same as sheep and cattle.[74]

At first, Finley was certain that Darling agreed that grazing should be forbidden on refuges, and he hoped to get Darling to take the matter up with Ickes. He wrote to H. M. Worcester, warden at the Clear Lake Refuge (Klamath),

> Since this matter of grazing is going to be carried up direct to Secretary Ickes, we want to get a definite understanding and I feel that it may be best to work toward the elimination of all grazing if possible on wild fowl reservations. I spent an afternoon recently with Senator McNary, and he says that the whole thing will depend upon Ickes. Darling has Secretary Wallace interested, and they are going to fight to get all they can.

Finley eventually came to realize that, for Darling, fighting grazing was not worth the political costs of battle, and he wrote to Worcester in frustration,

> I feel that it is a mistake for the Survey to sanction in any way the grazing of stock on the reservations. I have, therefore, written a long letter to Darling telling him I just returned from Tule Lake Refuge, and telling him the conditions. . . . However, he has a thousand and one things to attend to and must leave a great many of them to men under him unless special attention is called.[75]

For all of Finley's efforts, grazing and haying continued on the refuges. Even he recognized that, given the ongoing legal issues of land ownership, allowing haying to continue might be politically advisable. He wrote in 1935, "In the later summer and early fall after the nesting season, a large amount of wild hay can be cut around the border. . . . Malheur Lake Refuge will prove of real value to residents of [Harney County] when the government has completed its plan."[76] Initial grazing and haying programs on the refuge had been designed partly to raise funds for restoration projects, as well as to lessen local hostility to the refuge. Finley agreed to grazing leases only with great difficulty, for he felt the federal government had reneged on its promise to fund Malheur's restoration work. Unlike many managers, he hated the idea that refuges should be made to "work" for their living, since he believed that the "natural" work they did was more valuable than the few receipts received from grazing and haying. But he conceded the need for a small grazing program, one that soon mushroomed into a substantial source of revenue—and controversy—for the refuge.

Whereas Jewett, Finley, Benson, Cantwell, Cattron, and Triska had been reluctant to allow haying or grazing on the refuge, Scharff wholeheartedly encouraged it. Having grown up on a ranch, he believed that grazing was a good thing— good for livestock, ducks, relationships with locals, and money to fund water control work. In his first annual report, Scharff featured a photograph of cattle standing about in the uplands (plate 16). The ground is visibly bare, dusty, and very heavily grazed. Underneath this image, the caption reads, "A Source of Revenue for Years to Come." In fiscal year 1936–37 Scharff added that over thirty thousand AUMs (animal unit months—a cow and her calf foraging for one month) were using the refuge, while nearly six thousand tons of hay were cut and stacked on refuge land. More important, he urged full use of all refuge grazing resources, worrying that too little was being done on a national level to promote grazing on refuges. He bragged that under his watch, "no opportunity has been avoided" to promote grazing.[77]

The next year, Scharff wrote in his narrative report that "economic use" permits had dramatically increased, although "we have been unable to swell the economic use revenue as much as we would like to do."[78] In his 1939 report, he quoted from the Oregon Cattleman, a stockholder's journal: "The Malheur Refuge is just what the word implies and a large number of stock have been taken from the ranges to the lake which should prove beneficial both to cattle and to the range."[79]

Why such an emphasis on use? Partly, Scharff wanted the money for water control. And partly, he was a former rancher who still valued business stability and full use. In his 1938 report he noted that "5 year permits were issued April 1 to stabilize business."[80] In 1939 he complained that only "40 percent of the Refuge's feed palatable to livestock was utilized," a situation he perceived as terrible waste. In 1947 he warned that the refuge livestock program was still "nowhere near carrying capacity," and noted that cattle were "gold on the hoof in refuge meadows."[81] In that 1947 report, signs of trouble with overgrazing were beginning to creep into Scharff's prose: "During the year, the refuge pastures all held up well and with minor exceptions an abundance of palatable forage remained at the close of the grazing season."[82] Pictures in the report show abundant signs of overgrazing: few willows along the ditches, riprap work on dikes, and banks completely denuded of grass and shrubs.

Scharff saw no conflict between cattle and ducks, and local ranchers agreed with him. His grazing program did achieve many of its goals: cattle grazing benefited ducks during certain parts of their nesting cycle, grazing increased revenues for water-control programs, and leases reconciled local opinion to the refuge. Scharff was extremely popular among ranchers, and ducks were easy to see on grazed refuge lands. Cattle numbers continued to climb, until by 1968, cattle AUMs had reached over 137,000 for that year alone—livestock use nearly as intense as during the days of the cattle barons.

CARP CONTROL

While Scharff was happily maximizing cattle production on the refuge, he was finding other aspects of Malheur's nature much more difficult to manage. One of the most profound and unwanted changes to the Malheur water system came when carp were introduced in the Silvies River, most likely by pioneers during the late nineteenth century as a food source.[83] Few people proved to like the taste of carp, however, and carp populations soon exploded, with a host of unintended effects. Carp made their way from the Silvies River into Malheur Lake, perhaps during the high water year of 1952, when floods may have flushed many fish into the lake.[84] Bottom feeders, carp churned up sediments and destroyed sago pondweed. Because sago pondweed was a critical food source for waterfowl, duck populations plummeted at Malheur. By 1955 sago pondweed was almost gone from Malheur Lake, and by 1957 carp had made their way up the Blitzen nearly as far as Page Dam, some forty miles upstream. This unruly bit of nature—an unnatural introduction but profoundly natural in its unwillingness to abide by human rules—became a profound threat to the empire of nature at Malheur.

The refuge responded by initiating a series of poisoning projects whose intensity and scope were made possible by technological advances that had resulted from World War II and by a worldview that had declared war on any aspects of nature that refused to accede to human control. Refuge staff set out to control carp by dumping and spraying the fish poison rotenone throughout the system—an enormous project, for it involved poisoning the Blitzen River, the Silvies River, all their tributaries, and the lake itself. During several dry years the lake levels had dropped quite low, shrinking the lake surface.[85] With an extensive dike that stretched across part of the lake, and with water control structures along the Blitzen River, the staff shrank the lake even further, making carp control feasible.

In the fall of 1955, the killing began. The quarterly report detailed the elaborate process:

Toxicant was placed in the Blitzen River beginning on October 15 at Page Dam, approximately 40 miles above the mouth. The material was introduced directly into the stream from 55–gallon steel drums on the basis of 2½ pints of toxicant per acre-foot of water passing a given point. Additional drums were placed at Grain Camp Dam and near refuge headquarters. Known as drip stations, these drums poured toxicant into the river at a constant rate. The current carried the material through the entire course of the stream below Page Dam. Toxic water began entering the lake the evening of October 16. On October 17, the lake was sprayed by a contractor using flagmen and a converted TBF navy torpedo bomber which carried 640 gallons of toxicant, which was used without dilution. . . . This dosage was considerably above that recommended by the manufacturer. . . . The various sloughs, lakes and marshes of the Blitzen Valley were treated by a small aircraft or by hand back pump cans.[86]

Photos show that staff were knee deep, hip deep, neck deep in dying fish. "This carp mortality in the Blitzen River continued until October 19, and was so extensive that jams of dead fish were formed on brush at some points."[87] The reek was extraordinary.

At first, staff were delighted, writing in 1955 that "results, as a whole, from the spraying of the Blitzen Valley and Malheur Lake were gratifying. An estimated 1½ million carp were lying dead on Malheur Lake."[88] The Carp Control Project killed a lot of carp, but not enough. Although staffers were at first certain they had won the war against trash fish, by the next spring they were much more sober, writing that "approximately 2,000 adult carp had survived the 1955 treatment of the lake and apparently spawned successfully."[89]

Within just a few years, carp seemed as numerous as ever, numbering in the millions just several years after the killing. Control projects continued for several decades. Other aerial sprayings were undertaken during low-water years, with equally limited success. Drawdowns of water to dry out carp-filled ponds, as well as spraying, continued in sporadic but helpless attempts to wipe out carp.[90] Eventually, by the 1970s, staff admitted defeat and gave up on Malheur Lake as prime waterfowl habitat, concentrating on the Blitzen River instead, where they could have some hope of controlling carp.

Not only carp died in these control programs, of course, for rotenone is deadly to most fish and kills many of them more quickly and thoroughly than it kills carp. The Carp Control Project report noted, "The effects of the toxicant in the upper treated portion of the main Blitzen River were evident within one hour from the time of introduction, when a number of rainbow trout, whitefish, small shiners, dace, squawfish, and suckers were killed. The reaction of carp to the toxicant was much slower, as expected."[91] What about all these other species? From today's perspective, one immediately wonders under what moral or scientific calculus one can kill all aquatic species, just to increase food for ducks. What about the native fish in the basin?

Surprisingly enough, very little was known about native fishes in the basin when the 1955 Carp Control Project was initiated, even though the Oregon State Game Commission and the Fish and Wildlife Service had been stocking the basin for years with non-native game fish. The project report stated, "The only list of fishes indigenous to Harney Basin available before the project was undertaken was contained in the unpublished correspondence between Stanley G. Jewett, Jr. and Carl L. Hubbs."[92]

The refuge biologists, for all their lack of knowledge about native fish populations, did express some concern that extinctions might result from the Carp Control Project. This concern, however, did not stop them from proceeding. After the project's completion, staff argued, "It is believed that none of the indigenous species were eradicated by the project because specimens of all (with the possible exception of the coarse-scaled sucker) of the known species were found upstream from the areas treated for carp."[93] Knowing little about native fish biology, the refuge staff failed to realize that breaking the connections between lake and river populations of a species could result in extinctions, as the next chapter explains in more detail.

Eradicating native fish was nothing new. In 1950 Scharff had authorized replacing them in refuge waters with sport fish to make the refuge more attractive to sport anglers. He reported,

In 1952 and 1953 Boca Lake was rehabilitated with rotenone by Service biologists and stocked with rainbow trout in an effort to establish a source of spring-spawning rainbow trout eggs. The experiment failed to produce the desired results. However, the stocking program for the Blitzen River has produced excellent sport fishing for rainbow trout which grow up to 20 inches in length in these waters.[94]

This description of the lake as "rehabilitated," as if native fish species were criminal, is linguistically akin to the Bureau of Reclamation's use of "reclamation" to mean the eradication of native ecosystems.

What decades of ranchers and drainage efforts had failed to do, carp managed quite nicely: they transformed Malheur Lake from splendid water bird habitat to something still magnificent but far less productive for birds. Carp had inadvertently created another nature—a monster to some, a place of incredible fecundity to others.[95] Nature kept recreating itself, a many-headed hydra, escaping from the bounds people attempted to place upon it. People were responsible for these monsters, but they had little luck controlling them. Scharff was not troubled—perhaps few people were in those post–World War II days—by the prospects of failure. Torpedo bombers, undiluted poisons, hacking the heads off millions of carp: anything was possible in the war to create a better nature. Eventually, however, after Scharff's retirement, the refuge staff admitted defeat in the struggle against carp and focused instead on merely keeping carp populations from exploding to the point that they displaced everything else in the marsh. They realized that, instead of complete control, they would have to find an uneasy relationship of give and take with the new nature they had helped to create.

These were the glory days of the refuge—the days of water control, grazing, haying, spraying, carp control, willow eradication, beaver killing. In 1951 local historian George Brimlow described the engineering in terms that evoke the enthusiasm of the age for such projects:

> Engineering feats accomplished wonders for the sanctuary of migratory waterfowl and the haven of feathered and furry life. A concrete dam spans the Blitzen at the south end of the refuge. . . . The Blitzen serves as a canal for twenty eight miles, there being 100 miles of canal. Water is held at various levels, the impounding of it in a box canyon making Boco Lake. Dikes and other improvements assure proper control and protection.[96]

The Fish and Wildlife Service hoped to construct an empire of nature, a complex machine aimed at increasing waterfowl production. But trying to maintain this "proper control and protection" led refuge staff into continued complications.

The seeds of empire had been planted in the 1920s and 1930s by Finley, Gabrielson, and Jewett—preservationists who felt they could challenge the reclamation empire with an empire of nature based on a dream as potent and moving as the romantic myths of the Wild West and cattle barons. But their empire of nature became a troubling, complicated dream that rested on assumptions as shaky as the dikes and levees. Just as the emphasis on controlling water grew out of the reclamation conflicts and mushroomed into something much more complex and contradictory, so too did the grazing and haying programs grow out of a reasonable strategy to lessen local hostilities. Shifting political strategies and alliances between human groups had powerful effects on the shifting boundaries between water and land. What happened to riparian areas in the West stemmed from scientific ideals, engineering assumptions, the perceived need to control water—and also from the need to reconcile tensions among different cultures in the valley.

5 / Grazing, Floods, and Fish

F or nearly five decades ranchers and refuge managers had, for all their conflicts, found common ground in the promises and hopes of progressive land management. Both groups believed they could reshape the landscape to increase production, thus engineering a brave new nature. Both groups believed that through water manipulation, the replacement of native riparian vegetation with exotics, the removal of competing animals and plants, and other forms of intensive management, they could have both cattle and ducks.

In the 1970s this faith began to crumble. Growing concern over grazing effects on waterbird nesting soon turned into concern over larger issues of riparian health. When refuge managers began trying to restore riparian areas by lowering cattle numbers, a firestorm of controversy erupted over cows versus birds, swamps versus farms. The terms of the debate had shifted since the conflicts of the 1920s and 1930s, but many of the basic assumptions about nature that fueled the debates had changed little in the intervening decades. The fault lines laid during the first three decades of the twentieth century were still profound at the century's end.

GRAZING

The first rumbling of trouble on the refuge came with the grazing program. Denzel and Nancy Ferguson were two biologists working at Malheur Field Station, a research and teaching facility located on refuge property but supported by a consortium of universities rather than by the Fish and Wildlife Service. The Fergusons began to protest cattle grazing on the refuge, and in 1983 they published *Sacred Cows at the Public Trough,* a book that called the nation's attention to grazing excesses on public lands. As the Fergusons recounted, visitors to Malheur

were greeted with scenes of total devastation—cow manure as far as the eye could see, emergent vegetation trampled and reduced to filthy mats, dead cattle and cattle skeletons littering refuge waterways and canals, cattle wading and urinating in streams soon to be opened to public fishing, trenches worn several feet deep into ditch banks and levees, willow thickets broken and scattered, and vegetation eaten to bare ground. All this for the benefit of about 60 ranchers, many of whom owned airplanes and drove Cadillacs."[1]

Not surprisingly, the Fergusons angered many of those ranchers—they were tossed out of local dances, had their lives threatened, and finally were ostracized from the community.[2] The Fergusons argued that between 1948 and 1974, livestock permits at Malheur had doubled, whereas the duck population had dropped seven-fold, from 151,000 to 21,300. Although the number of waterfowl that use the refuge has always fluctuated, the Fergusons attributed these declines to grazing rather than to fluctuations in climate, rainfall, carp, or changing conditions elsewhere along the Pacific Flyway.[3] Simple correlations between cattle and ducks were appealing to those environmentalists who hated grazing, but they were problematic. Correlations could as easily support the opposite argument. After John Scharff retired and the Fish and Wildlife Service began slowly reducing cattle numbers, locals responded by attributing declines in nesting birds to the cattle reductions. In *Surviving the Second Civil War,* the rancher Timothy Walters wrote that during the 1950s, when cattle numbers were high at Malheur, the refuge had the highest nesting population of sandhill cranes in the country and was host to tens of thousands of waterfowl. But after grazing permits were cut, "only 2 greater sandhill crane chicks were raised at Malheur during the 1973–74 season" and wildlife migrated from the refuge to nearby ranches, where "the food supply was still intact."[4] The refuge typically produces far more waterbirds than surrounding private land produces, but since during spring migration, more birds often use private lands, locals assumed the reduced grazing on the refuge was hurting birds. Waterbird numbers are controlled by ecological factors far more complex than simple numbers of cattle, but both environmentalists and ranchers focused on grazing.

As Scharff continued to increase grazing on the refuge, outside Fish and Wildlife Service biologists became increasingly concerned about both the bad publicity and the effects of cattle on marsh habitat. When Scharff retired, biologist Joe Mazzoni was brought in as refuge manager and given the difficult task of restoring habitat by reducing cattle numbers. Agency staff knew this would be an unpopular change in policy, and local ranchers proved them right when they responded with an explosion of public anger to cattle reduction. One local man named William D. Cramer wrote a letter to the *Burns Times-Herald* in 1978 that called Mazzoni an "extremist" trying to sell "propaganda" to a "gullible public." Cramer wrote,

Whoopee! Joe Mazzoni says his devastating policy of leaving the meadows unmowed is having great results. There has been a substantial increase of small rodents and a corresponding increase in birds of prey. The theory seems to be that birds of prey are better than no birds at all on the Malheur Refuge. Mice and hawks and Canadian thistles. We should all be proud of what these great thinkers are doing with our public lands. How long will Congress allow this type of nonsensical experimentation to continue?"[5]

For all his sarcastic anger, Cramer had noted a critical shift in refuge policy. The refuge was no longer trying to maximize waterbird production at the expense of all other forms of biodiversity. Not surprisingly, this shift did not appeal to locals, many of whom saw hawks and rodents as the very worst of nature, something no decent American should tolerate, much less help.

In his first year at the refuge, Mazzoni actually increased livestock AUMs: from 97,900 in 1973 (Scharff's last year) to 98,500 in 1974. However, 1973 had been a drought year, so Mazzoni had actually reduced grazing 16 percent from the peak years. He did not describe his program as the first steps in getting cattle off the refuge. Instead, he used far more politic and bureaucratic language, explaining his reductions as

our continuing effort to develop the kind of flexibility in our vegetation management program needed to improve the quality of waterfowl nesting cover and eliminate (or at least minimize) some of the traditional conflicts we've had between our grazing and haying program and successful greater sandhill crane production.[6]

And indeed, the intention was not to eliminate grazing—as the Fergusons wanted and early refuge advocates had desired—but to reduce it to what managers felt was a reasonable level, one that allowed extensive use by locals but lessened pressures on other animals. Cattle were still conceived of as fundamental to the refuge program, even though biologists realized they competed with native animals and plants.

In the margins of copies of the refuge managers' annual reports sent out for review, Fish and Wildlife Service staff in Portland scribbled comments revealing that opinions about grazing had begun to shift throughout the region, not just at Malheur. One comment noted that the number of AUMs in 1974 was still "quite a ways from the 50–60,000 desired. Hang in there." Another commented, in response to the terrible local publicity Mazzoni was receiving, "I see your friends (?) are giving you some help." Yet another wrote, "Stay with it, Joe!" Fellow Fish and Wildlife Service biologists recognized how difficult Mazzoni's work was, no matter how slowly he had tried to institute reductions.

When Mazzoni began reducing livestock on the refuge, depressed cattle markets throughout the West magnified the fear many local ranchers felt. Mazzoni wrote in his 1975 report,

> One product of the depressed livestock market experienced this year was a stronger than normal rancher resistance to any change in federal land management programs that might adversely affect their operations. Approximately 92 percent of Harney County's agricultural income is derived from cattle . . . and gross farm receipts were reduced to almost one-half of last year's gross due to low cattle prices. . . . This had an important bearing on rancher attitudes towards changes being implemented in our program, and tended to complicate our working relationship with them."[7]

Rather than reducing the AUMs of existing permit holders, Mazzoni tried to lessen local anger by retiring existing grazing permits when ranchers sold their lands. This strategy allowed him to reduce AUMs almost 50 percent by 1977.[8] In 1972, under Scharff's watch, only 650 acres of nesting habitat had been free from grazing or haying operations. By 1977 that figure had increased to forty-two thousand acres of marshland.

The damage done to riparian areas at Malheur was not a problem limited to wildlife refuges. Ironically, programs designed to help restore grazing damage on federal lands throughout the West may have actually helped accelerate damages to riparian areas. Horrified by the effects of overgrazing on uplands and mountain pastures, the early ecologist Arthur Sampson had promoted a plan in the early twentieth century to restore grasslands through what he called "rest-rotation grazing." Under this system, grazing lands would be fenced into a two- or three-pasture system, which would allow each fenced pasture to receive a year of rest from grazing. Some years a pasture would be grazed early in the season; other years it would be grazed later. The intent was to allow plants a year of rest to recover vigor, and in the late-grazed pastures, to enable plants to set seed before being grazed. Pastures would be grazed not until all available forage was gone but until about 75 to 90 percent had been utilized.

For upland grasses and forbs, the rest-rotation system could work well, allowing recovery of some bunchgrasses. But utilization standards ignored the fact that cattle did not distribute themselves evenly over pastures. Instead, they tended to congregate in riparian areas, moving into the uplands only after their favored riparian plants were eaten. Waiting to move the cattle until after they had eaten most of the upland plants did nothing to protect riparian plants. Moreover, riparian shrubs, willows, cottonwoods, aspen, and sedges could not prosper with only a year of rest, or with late-season grazing. Rest-rotation grazing, promoted by government specialists and land-grant institutions as the modern, conservation-minded way to graze and restore grasslands, helped devastate riparian areas. As the riparian specialist Bob Ohmart wrote, "Up until the late 1960s, riparian habitats were viewed as sacrifice areas" by federal land managers.[9]

But why did grazing specialists not notice the damage being done to riparian areas? Riparian specialists Wayne Elmore and Bob Beschta have argued that problems in riparian management have developed in large part because of human perceptions, not just because of the presence of cows. Since degradation of many riparian areas happened early in the century, many current landowners and managers, who have not seen land in an undegraded state, believe that current riparian condition is "natural." Elmore and Beschta found that few ranchers or managers had ever seen a "healthy" riparian area."[10] Many ranchers believed, and still believe, that creeks are not supposed to be covered with willows; they are supposed to be bare, to allow cattle easier access to water. As one rancher wrote, "we were brought up to think riparian areas were supposed to look well maintained by cattle."[11] What ecologists have learned to see as a sign of a damaged creek, others still see as normal. Those same ranchers might well prefer the benefits that come with a healthy, vegetated riparian area— more abundant forage and water—but they do not see those as connected in any way with the creek bank. People's conception of the Great Basin, according to the ecologist D. Dobkin, has been distorted by over a century of livestock grazing: "We have lost from our collective consciousness what these landscapes looked like before fire suppression and grazing in the Intermountain West."[12]

Riparian areas have suffered not just from perceptions based on a faulty reading of history but also from a certain invisibility. Because riparian areas are neither water nor land, they do not fit into normal categories of analysis. To many people, these areas do not exist, and so they are overlooked. Even in the 1990s, few cattle ranchers or locals considered streamside areas important or worthy of attention. For example, Jack Southworth, a cattle rancher near Seneca in the northern end of the Malheur Lake Basin, spent years managing his ranch without paying attention to the riparian areas. In his words,

I used to give a class at Oregon State University about ranch management. I had a slide that showed some nice cross-bred cattle. I showed this slide for three years before I saw the creek. The creek banks were really eroded, and I didn't see it until someone in the class pointed it out to me. What's happening throughout our society is that we ranchers for all time didn't see the creek, and now we're starting to."[13]

Southworth remembered that for decades his family

> did everything we could to simplify it [the ecosystem] to make it easier to understand, and therefore to manage. For example, in the 1950s, the Soil Conservation Service built a nifty plan to straighten the Silvies River where it runs through a couple of miles of our meadows. And we got rid of those obnoxious willows that got in our way during haying; those branches would break a mower section now and then. But, then we had to build rip-rap along the stream banks to stabilize them. And the more we managed it, the worse it became.

The change in perspective described by Southworth happened to refuge managers as well as to some ranchers. With John Scharff's departure, riparian areas began to receive attention on the refuge. In 1977 Mazzoni made what appears to be the first specific mention of riparian areas in four decades of refuge reports. Mazzoni wrote that along Krumbo Creek, "intensive grazing use had virtually eliminated streamside tree and shrub species (other than sagebrush)" while also accelerating erosion.[14] By 1986 refuge reports were describing attempts to balance tradeoffs between riparian health and water control:

> Management objectives of woody riparian zones are aimed at minimizing conflicts between different wildlife species needs. For example, the uncontrolled spread of a willow patch on a delivery canal will provide improved habitat conditions for willow flycatchers. However, the decreased water control capabilities associated with a willow filled ditch could cause down stream meadows and marsh to dry up and become less valuable as waterfowl habitat."[15]

Refuge managers had faced these tradeoffs ever since management began, but before Mazzoni's era, managers did not even seem to question whether better water control was always worth degradation of riparian habitat. By 1987 the new refuge manager George Constantino could write that riparian zones were being actively restored, not just tolerated: "Woody riparian zones are managed to provide optimum conditions for a multitude of species including red-band trout, willow flycatchers and other songbirds. Management included the efforts to protect these areas from disturbance and promote healing of areas previously disturbed."[16] Not surprisingly, many local ranchers and residents thought these changes in goals were absurd. Willow flycatchers? Hawks? Had the refuge staff gone mad? Had they been taken over by extreme environmentalists?

Tensions finally exploded in a bizarre echo of the attack on Peter French when a local rancher, Dwight Hammond, assaulted refuge staff in 1994. Hammond had been making death threats against refuge managers for years: against Constantino in 1986 and 1988, and against Forrest Cameron, the next manager, in 1991. Hammond had repeatedly violated the terms of his refuge permits, abusing his right-of-way privileges by refusing to give the required notice before moving his herds across the refuge. He often allowed his cattle to trespass for days on refuge streams, trampling willows that restoration crews had just planted. In frustration, Cameron notified Hammond that the refuge was going to build a fence along the refuge boundary with Hammond's land to keep the cows out of the canals and streams. On August 3, 1994, a crew came to put up the fence, but Hammond had parked his Caterpillar on the boundary line, first removing the battery so that the refuge crew couldn't move it out of the way. When the refuge called a tow truck, a bizarre showdown ensued. Hammond hopped up onto the Caterpillar and lowered the bucket, "narrowly missing another special agent," while his son Steve shouted obscenities at federal officials.[17]

Nine federal agents took the Hammonds into custody, and the men were charged with "disturbing and interfering with" federal officials—a felony. The Hammonds spent two nights in jail, and an already tense situation soon led to demonstrations by Wise Use antienvironmentalists and threats against refuge biologists. As one news report detailed,

> On Aug. 10, nearly 500 incensed ranchers showed up at a rally in Burns featuring wise-use speaker Chuck Cushman of the American Land Rights Association, formerly the National Inholders Association. Cushman later issued a fax alert urging Hammond's supporters to flood refuge employees with protest calls. Some employees reported getting threatening calls at home.

Wise Use leader Chuck Cushman threatened to "print a poster with the names and photos of federal agents and refuge managers involved in the arrest and distribute it nationally. 'We have no way to fight back other than to make them pariahs in their community.'" Another Wise Use group declared, "This is a hostage situation!" The Republican congressional representative for the district, Bob Smith, wrote a nasty letter to Interior Secretary Bruce Babbitt, attacking the actions of the refuge staff (and incorrectly accusing the feds of arresting Hammond at his own house, rather than on refuge property, where the arrest actually took place). In response, the U.S. attorney's office reduced the charges and indefinitely delayed the hearing, infuriating environmentalists and delighting the burgeoning Wise Use movement in eastern Oregon.[18] As Forrest Cameron said in an interview, Hammond had initiated a battle over "who was going to run the refuge."[19]

Conflicts among environmentalists, ranchers, refuge managers, and Wise Use groups revealed how public grazing policy on the refuge had become. Instead of a policy decided quietly by John Scharff and his local buddies, grazing at Malheur had become a national controversy. These conflicts, intense as they were, were soon to be overshadowed by far greater battles over floods and sensitive native fish.

FLOODS

Although political forces contributed to changing policy on the refuge and throughout the basin, the most radical changes of all were wrought by natural forces: floodwaters in the early 1980s that reshaped the face of the ecological and human landscape. Heavy snowfalls led to increasing water levels each spring, until by 1984 floodwaters filled much of the closed basin, wiping out farms one after another, washing out roads, ripping out culverts, and calling into question the way the world was structured. What shifted with the waters were not just the boundaries between water and land, but cultural attitudes. These floods, coming hand in hand with battles over grazing leases, undermined the post–World War II belief that riparian landscapes could be reshaped into orderly agricultural machines.

To understand the effects of the floods in the 1980s, it helps to step back and review local efforts to develop water resources and control floods. Floods were nothing new: they had recurred for millennia. Ecologists had begun to argue in the early 1980s that floods were a critical element in the functioning of riparian landscapes. Yet, although floods were certainly natural, the effects they had on the basin had changed dramatically in the past century. Before white settlement, many of the most damaging effects of floods had been buffered by abundant riparian vegetation. Riparian plants had increased resistance to high water flows, slowing the speed of floods and reducing the erosive power of floodwaters. Side channels, meanders, beaver dams, debris in the creeks, and the sinuous, swampy landscape had all worked to moderate the impacts of floods. But farming, grazing, and channelization of the Silvies and Blitzen Rivers had radically changed the basin's response to flooding.

The same human modifications that had reduced the ability of the basin to absorb flooding also made people less willing to live with floods. Early ranchers who had flood-irrigated wild hay meadows had been relatively willing to live with the inconvenience of annual floods. Benefits from flooding were clear: water, lush grass growth, subirrigation, and sediment deposition that increased the quality of basin soils. The ranchers recognized that if floods were cut off from their meadows, sagebrush and other upland vegetation would move in, reducing ranch income. Early ranchers protected their houses and barns from the spring waters, and they tried to manipulate where the floodwaters ran in the spring, but they otherwise had been willing to adapt to the floods, living with both the costs and the benefits.

But as ranchers made the switch from wild hay to alfalfa in the 1950s, effectively turning their holdings into what Scharff called "beef factories," they were less willing to adapt to floods. John Scharff described this transformation approvingly: "The income is rising spectacularly on many ranches as the owners fertilize the meadows, replace sagebrush with grass, use water to better advantage, get higher calf crops, and perfect the hundreds of management factors that make a ranch a better beef factory."[20] To many locals who were struggling to "perfect" their beef factories, the problem seemed clear. Floods washed over the Silvies Valley early in the spring, making it impossible to get the heavy equipment needed for planting alfalfa into fields, yet in the late summer, water needed for irrigating alfalfa ran out. Why not simply build a storage reservoir that would hold back the waters in the spring and release them in the late summer? Even though Reclamation Service officials in the early twentieth century had spent many years writing reports that declared such projects unfeasible, the political landscape changed when the Army Corps of Engineers came into the picture.

THE 1957 ARMY CORPS REPORT

In 1941 the Flood Control Act authorized the Army Corps of Engineers to survey rivers across the nation for flood control.[21] In 1945 the district engineer of the U.S. Engineer Office produced the "Report on Preliminary Examination for Flood Control of Silvies River and Tributaries, Oregon," which recommended an extensive survey for flood control. In 1957 the Army Corps of Engineers finally published the results from this survey, arguing that local water conflicts had become acute because very little water from the Silvies ever reached Malheur Lake. Most was used in irrigation, and much of the rest was lost to percolation or evaporation. But although irrigation water was running short, the corps wrote in 1957 that "irrigation practices are showing improvement on many of the ranches" and that primitive flood irrigation was giving way to more efficient projects: "Large earth-moving equipment has been brought into the basin and is available at reasonable cost for constructing levees, clearing or reconstructing canals, and similar operations."[22] Bulldozers had begun to allow extensive and relatively cheap channelization of local streams and waterways, not to mention the construction of levees, which would keep floodwater off the land.

The 1957 army report made the case for a storage and flood-control reservoir, arguing that floods threatened efficient ranching and "urban living" (although it was a stretch of the imagination to call life in the little ranch town of Burns "urban"). From the perspective of army engineers, the "destructive effects of annual flooding" made it difficult to grow "the better types of hay, generally limiting crops to native grasses." Because of these floods, "diversion and distribution systems are generally rudimentary and little attempt has been made for refinement of the irrigation system or improvement of the natural channels. As a result, flood damages are aggravated by presently necessary irrigation operations."[23] Army engineers complained that ranchers still growing wild hay in their meadows were building primitive dikes across little natural sloughs and backwaters that had always filled with floodwaters during the spring runoff. The structures kept slough waters from flowing back into the river after floodwaters receded, thus providing ranchers with "primitive" storage reservoirs, while also creating new riparian habitat. But these small structures exacerbated the effects of flooding on neighboring lands where modern ranchers were trying to raise alfalfa: "These structures cause direct overflow and flooding of large areas of adjoining land, to an extent not necessary for irrigation."[24]

This argument contains two revealing elements. First, floods were thought to have kept the region in a state of backwardness, for they limited the construction of extensive engineering projects for modern irrigation. No one wanted to develop expensive improvements without more flood control to protect their investments. Second, it was evident that social relationships within the valley had complicated the effects of natural relationships. The ranchers who continued to use primitive flood irrigation worsened the effects of floods on their neighbor's lands. In its natural state, the basin would have been less vulnerable to the natural effects of flooding but more vulnerable to engineering transformations. Relatively minor and inefficient manipulations of the landscape had interfered with the much more extensive manipulations that modern engineers and modern ranchers desired.

The army report argued that floods were being made worse by "lack of adequate natural channel capacity."[25] This seemingly innocuous statement assumed that channels were not incised or damaged enough. If only they were even deeper and wider and more degraded—more like ditches and less like rivers—they could contain the floods, engineers believed. Overbank flooding—which ecologists and hydrologists now see as the major benefit of high water—was considered the major problem by army engineers. The incised channels—which ecologists now see as the central problem—was considered the solution.

The corps recommended construction of a storage reservoir in the Silvies River canyon, the same place that earlier reservoirs had been proposed. Oregon

State University's Agricultural Research Station and the School of Agriculture strongly favored the project "because of the anticipated benefits to the basin from improved utilization of an available water supply." The refuge, the state Game Commission, and national conservation groups opposed the reservoir plan, arguing that the refuge had legal rights to floodwaters for the lake. The corps had calculated that the proposed reservoir would reduce lake levels by about one foot and reduce lake area by an average of about eleven thousand acres. The refuge argued that these effects would harm critical habitat, but the army appeared unconcerned, barely mentioning the refuge in the report. The response of army engineers to the problem of siltation—an effect magnified by destruction of riparian habitat—was particularly revealing. Engineers did calculate possible siltation rates, and figured that sediment would fill in "60 percent of the proposed dead storage over a 50–year period," which certainly sounds significant to a reader today. Army engineers, however, dismissed their own findings, arguing that "the silting of the reservoir would not become a serious problem for many years."[26]

Although seemingly unconcerned about ecological effects, the army was concerned about economic effects. To decide if the project could be justified economically, the army calculated a benefit-cost ratio. Such ratios were supposed to be a quantitative, and therefore unbiased, way of measuring the value of a project. Army planners estimated the total cost of construction at $5,454,000, with annual costs of $292,000. Annual benefits were estimated at $408,700, largely because of potential flood control benefits to the town (rather than to rangeland), in addition to potential irrigation benefits. The report claimed that the benefit-cost ratio of 1.4:1 showed that the project would be economically beneficial for the basin.[27] The project did not go forward, however, because contested water rights derailed it, just as they had derailed every other reclamation project ever proposed for the basin.

Benefit-cost ratios were a tool favored by planners because they were seen as value-free and therefore beyond attack. Numbers, rather than values or political alliances, surely offered a rational way of making decisions, planners believed. But values shaped the assumptions behind these calculations, and as values changed, the calculated benefit-cost ratio could also dramatically change.

Twenty years later, in 1977, the Army Corps carried out another study on the same project in the same location, with new benefit-cost ratios that revealed fundamental changes in values.[28] This time around, estimated project costs were calculated to be over $19 million, 3.5 times the cost projection twenty years earlier, far outpacing inflation. The annual projected charges for the reservoir increased from \$292,000 in 1957 to \$3.5 million (an increase explained largely by the need to service debt from initial construction costs). Most strikingly, total annual benefits decreased to \$395,000, so the benefit-cost ratio dropped to 0.11:1.[29] The report stated, "In view of the obvious lack of economic feasibility, it is recommended that this study be terminated and no action taken toward broad authorization of a project at this time."[30]

In part, these calculated benefits changed because engineers no longer assumed that flood control would really benefit the town. In 1957 the army had written dramatically of possible devastation that flooding might cause to Burns.[31] In calculating the benefits of flood control in the 1950s, the army had relied on inflated projections of growth in Burns, estimating maximal damages to urban structures that did not yet exist. In 1977, after decades in which Burns had shrunk rather than grown, the army planners noted that 92 percent of the damage from floods in the past eighty years had occurred on cropland, not urban land. The report abandoned the optimistic projections of past planners, including only minimal urban flood-control benefits in its calculations.[32]

By 1977 the army had lost its optimism about the possible benefits from irrigation development. Rather than calculating the potential value of agriculture outputs from an entirely transformed basin, as the 1957 report had done, the 1977 calculations simply calculated the increase in value of irrigated hay. Planners therefore figured that the project would produce only \$52,000 in annual agricultural benefits from irrigation (the rest of the \$395,000 in benefits came from flood protection).[33]

The osu Agricultural Research Station stridently opposed the army's calculations, arguing that the project could create a new agricultural machine in the basin that would produce not \$52,000 in annual benefits, as the army calculated, but rather \$13 million. Their calculations revealed that a powerful optimism was still at work in the basin. Station staff calculated that by using every drop of the average flow of the Silvies River each year to irrigate alfalfa, farmers could produce 236,000 tons of alfalfa each year, worth at least \$45 a ton, for a total of more than \$10.6 million! Farmers could then convert each and every acre of marshland in the entire basin to alfalfa. Once water was kept off the land, farmers could use ground-water irrigation to convert sagebrush to improved pastures. Cattle on those irrigated uplands would surely gain over 700 pounds of weight per acre, so farmers could collectively produce 10,545,000 more pounds of beef per year, which they could sell for an extra \$2,636,000, leading to over \$13 million in annual irrigation benefits, not a mere \$52,000.[34]

The army responded by consulting with the Bureau of Reclamation, whose acting regional director gently pointed out some of the flaws in the station's projections:

The assumption that the entire runoff of the Silvies River, 118,000 acre feet, would be available for crop production (available to the plants) is not valid. The requirement for minimum streamflows, losses from evaporation, and conveyance losses both off and on the farm would reduce this amount by perhaps 40 percent.

Conversion costs to change wild hay meadows into irrigated fields would not be the $30 per acre figure offered by the station, but rather at least $1,150 per acre, and probably far more. In addition, the regional director asked, did the Agricultural Research Station really believe that all farmers could produce as much as the station's scientists produced on their experimental farms?[35]

The station's calculations existed in a dream world of imaginary agricultural perfection—a world in which John Scharff's fantasy of the landscape as a "better beef factory" could be realized. But the station superintendent did recognize one critical point: that the army's benefit-cost ratios reflected cultural assumptions. Essentially accusing the army of lying, he wrote in a bitter protest letter,

> Sociological pressures, pressures from others relying on water from the Silvies, and the reluctance of land owners to conform to the acreage limitation may be of greater impact on this project than it "not being economically feasible." Please do not confuse the real issues by inferring that the construction costs exceeded the expected increase in income unless it is a fact.[36]

Although the Army Corps in 1977 recommended against the project, this was not because environmental concerns had swept through the corps. From 1973 to 1977, Fish and Wildlife Service staff had written a series of angry memos to the Army Corps protesting the environmental consequences of the reservoir project. In the 1977 report, the army completely dismissed those concerns, merely noting that the reservoir might benefit wildlife in the Silvies River valley—a very odd conclusion.[37] Something had, however, changed. Army planners no longer assumed that their project could transform the riparian landscape into an agricultural machine of engineered perfection. Not everyone in the basin agreed, however; the fervent faith of the Agricultural Research Station staff shows that many people still believed in the ability of engineering to overcome natural constraints in the valley.

Although some local residents applauded the Army Corps' decision to shelve the project, others felt betrayed. Decades of study had resulted once again in the failure to construct a dam that could save the basin from its own natural vicissitudes. Just a few years after the flood-control project was turned down, the worst nightmares of planners and ranchers and farmers came true: floods that kept rising, year after year, until water had rewritten the story of the human and natural landscapes.

Heavy snows began in 1982 and continued for several winters, leading to slow but inexorable rises in lake levels. No wall of water came hurtling into town, no great floods of meltwater ripped through the basin, tugging children from their mothers' grasps. The waters rose slowly, spread slowly, and seeped slowly through the basin. By June 27, 1984, Malheur Lake had reached 4,102.4 feet: nearly 9.5 feet above its normal maximum level of 4,093 feet.[38] This may not sound like much, but in a flat basin, a little bit of water goes a very long way: each one-foot rise in lake levels submerged another 8,500 acres. The lake continued to slowly rise with each wet year that followed, until by 1986 it had reached 4,102.6 feet.[39] Whereas an average lake surface area of about forty-six thousand acres had once seemed "normal" for Malheur Lake, by June 1984 Malheur, Mud, and Harney Lakes had merged to cover more than 170,000 acres.[40]

The local economy, already hurting, suffered tremendously. After a rail-bed washout in March 1984, the Union Pacific Railroad closed its spur line between Burns and Ontario, Oregon. This line had been used to transport lumber from the sawmill in Burns, so lumber had to be trucked out, an expensive proposition.[41] Parts of two highways were submerged, along with fifty-seven thousand acres of marsh habitat. Twenty-five ranches were badly damaged, thirty families were displaced, power and telephone lines were destroyed, carp populations sky-rocketed to some 80 to 90 percent of fish biomass, groundwater was contaminated with arsenic, and fifty thousand acres of hay were lost.[42] Locals estimated the losses at $33 million, whereas the Army Corps estimated $21.6 million.[43]

Much to locals' anger, the State of Oregon and the federal government refused to declare the county a disaster area or to release emergency relief funds, largely because the slowly rising waters of a Great Basin flood did not meet outsiders' perceptions of what a flood should be: a sudden, sharp mess. The Bureau of Land Management and Malheur Refuge did their best to help. The refuge offered ranchers alternate grazing sites in the Blitzen Valley, and the Bureau of Land Management lowered grazing fees on public lands.[44]

The Army Corps was called in once again to solve the problem, and its planners and engineers went back to the drawing board. After several years of study (and several years of mounting frustration among locals) the corps proposed three possible solutions to the flooding: the oft-proposed Silvies Valley reservoir, an elaborate canal system that would drain Malheur Lake into the Columbia River system, and a nonstructural solution that entailed moving the railway line out of the floodplain and buying flood-prone lands from ranchers. The cost of the proposed reservoir had ballooned to $130 million (up from $19 million seven years before, largely because the army engineers now realized that a much bigger dam would be needed to control floods). The annual cost of the reservoir solution would therefore be $13,100,000, which even for army engineers seemed a bit steep in a basin where the most profitable crop was hay.[45]

The army figured that a much cheaper solution would be to drain water from Malheur Lake through a seventeen-mile-long canal into the Malheur River, which then flowed into the Snake River, then the Columbia River, and from there to the Pacific Ocean. The initial 1985 army reconnaissance report estimated the annual cost of the proposed canal at $4,400,000 and the annual worth of flood control at $2,160,000. Malheur Lake water could be sold to the hydroelectric dams on the Columbia River system, planners figured, and those annual power benefits would be worth $4,332,000, resulting in a benefit-cost ratio of 1.5:1. The 1985 reconnaissance study concluded that "the Virginia Valley Canal could be an effective solution to flooding around Malheur Lake and provide economic and social benefits."[46]

These estimates of costs and benefits, although seeming perfectly rational at first glance, proved insupportable even for the army. Most of the calculated annual benefits came from the sale of hydropower at dams on the Snake and Columbia Rivers. But, as an army report published two years later admitted, power benefits would occur mostly in the first two years, as drawdown occurred. After that they would be brief and intermittent.[47] Moreover, the army had estimated annual flood benefits of over $2 million. Critics quickly pointed out that several years earlier, in the 1977 report, the army itself had estimated annual flood reduction benefits at only $277,000.[48] If the worst flood in a hundred years had produced $1.2 million worth of crop damage, how could flood benefits be worth over $2 million a year? In a basin that produced such low-value crops, how could it be economically feasible to construct a project that cost far more each year than the crops were ever worth?

By the time the army completed the full report in 1987, its new calculations revealed much higher costs for a much smaller project. The revised canal would hold only half as much water, preventing only about 58 percent of flood damage in the future, and also reducing hydropower benefits. It would be those power benefits, not flood protection, that might make the plan economically feasible.[49] The army estimated a cost of $2.9 million to complete the project, and annual costs of $1,846,000. The calculated annual benefit totaled $1,683,000, but most of this derived from hypothetical hydropower production sales with little benefit to farmers.[50]

Economic calculations showed that the project would be impractical, but what about the environmental consequences? By 1987 the Army Corps of Engineers was required by law to detail possible environmental consequences. The report reveals that although planners did analyze possible effects, their own values prevented them from paying much attention to what those effects might mean for the ecosystems involved.

The primary environmental concern was that water quality in the Malheur River might suffer, eventually contaminating the Snake River and the Columbia River. Malheur Lake, as an inland sump basin, had the high levels of dissolved solids found in nearly all Great Basin lakes. Evaporation from Malheur Lake, as from most closed basins, concentrated dissolved solids, and the pH of the lake was usually above 7.0. Boron and arsenic concentrations were also high.[51]

What would that have meant for Malheur River? Water quality was very good in the upper watershed of Malheur River above Namorf Dam. Below, the quality was dramatically lower because of irrigation diversions and returns that concentrated dissolved solids and increased water temperature.[52] Namorf Dam was also an effective barrier to the carp that inhabited the lower river, so above the dam the trout fisheries were in excellent health—good enough to be one of the Oregon Department of Fish and Wildlife's "blue ribbon" trout fisheries. Native rainbow trout constituted 42.6 percent of fish sampled, the rest being hatchery trout.[53] Native Lahontan cutthroat trout (Oncorhynchus clarki henshawi) still thrived in the river above the dam, along with redband trout in some of the tributaries.[54]

If the proposed canal were to link Malheur Lake with Malheur River, this native trout fishery would be destroyed. The carp that filled Malheur Lake would find their way into the upper Malheur River, degrading the river as they had degraded the lake. The Oregon Department of Fish and Wildlife estimated that the annual fisheries losses would amount to at least $500,000, an amount not figured into any army calculations, even though the potential economic loss far exceeded the estimated benefits to farmers.[55]

The canal would have had other effects as well, of course. Releasing water from the lake into the river would have led to what the army acknowledged as "extensive erosion" along the Malheur River. The report proposed that "the most economical approach to erosion protection on these rivers was found to be a plan in which channel improvement and riprap construction prior to released water from Malheur Lake would be performed only in areas where significant damage would obviously occur." Nine miles of riverbank would eventually require bank protection and be "subject to high maintenance annually." In calculating these environmental costs, the army made no mention of the value of the lost riparian areas.[56]

The Fish and Wildlife Service did its best to draw the army's attention to the myriad environmental consequences of the canal solution. On April 10, 1985, the Division of Ecological Service sent an impassioned critique of the plan to Col. Robert B. Williams, the district engineer in Walla Walla, Washington. The critique pointed out some of the obvious economic and environmental problems with the plan: carp would ruin the Malheur River fisheries, dissolved solids and contaminates would lower water quality and harm irrigation in the Malheur River, sedimentation and erosion from the canal would damage the river's banks, and deer and antelope migration might be disrupted by the canal.[57] The army, however, decided that these were trivial concerns.

More telling, perhaps, than these lists of environmental problems, was the fact that the Fish and Wildlife Service argued that the very perception of flooding as problematic was at heart flawed, and that flooding was *not* an environmental disaster. Although flooding did eliminate much marshland, other marshland was created, leading to "an extensive increase in habitat outside of the flooded area" as "many playa areas and depressions in southeastern Oregon have flooded and developed into marsh habitat." Variability and flooding, although problematic for some animals and plants in the short run, were the true sources of the refuge's productivity, the critique insisted. The "historic drought/wet cycles in the marsh" had served "to recycle nutrients, aerate soils, set back plant succession. . . . The historic extremes of inflow and lake elevations have been a key factor in creating and maintaining wildlife values in the basin."[58]

These arguments stand in stark contrast to earlier attempts by the refuge to control water levels in the basin, as the Fish and Wildlife Service itself admitted. Although the service had once tried to regulate water levels in the marsh, new ideas about management that would "conform to natural water cycles" had recently played an important role in Fish and Wildlife Service management of Malheur Lake.[59] Or, as the 1986 annual report of the refuge argued, its new "philosophy has led to a management plan . . . that calls for natural hydrologic changes to continue unaffected by human intervention. . . . Most wildlife species have adapted very well to this natural change and the Service did not see any major losses occurring in the long-term."[60]

What led to this change in philosophy? Changes in the science of ecology, as scientists learned more about riparian and wetland function, may have contributed. Changes in people, as Scharff retired, also played a role. More fundamental answers to this question emerge in a 1969 article, "The Ecology of Malheur Lake and Management Implications," written by Harold F. Duebbert, a wildlife biologist with the Fish and Wildlife Service.

Duebbert recognized that some people might think "that manipulation in such a way as to perpetually maintain a peak level of productivity should be attempted." But then he pointed out new research that showed fluctuations to be critical for productivity in the long run:

> Perhaps one of the main factors in keeping shallow marshes productive is their seasonal and yearly fluctuation in water years. . . . Only when marsh bottoms are dry can necessary exchanges occur to maintain a proper balance between oxygen and other elements in the soil so necessary to the growth of marsh plants and animals.

He argued from a historical perspective that "it is no accident that the most productive marshes are those with a history of fluctuating water supply, while the least productive have stable water levels."[61]

Duebbert noted that although natural fluctuations were valuable for marshes, nearly all the marshes in the entire refuge system had already been regulated. Duebbert recognized that carp meant Malheur Lake was no longer pristine, but he believed that it still had value in its unregulated state. Across the nation's refuges, Duebbert complained, "almost universally, water was regulated by water control structures and dikes." John Scharff's 1965 refuge master plan proposed such a water control system for Malheur Lake, according to Duebbert.[62] Duebbert called this way of thinking into question: "The wisdom of disturbing the natural ecology of Malheur Lake with an intricate system of water control structures . . . is open to serious question." He went on to ask how artificial manipulation of water levels would affect the long-term ecology of the marsh. For example, would alkalinity increase? These were important concerns for him and for other wildlife biologists, if not for Scharff. But Duebbert recognized other concerns more profound than numbers, stressing what might be lost when human control had transformed all landscapes:

> A visit to Malheur Lake by anyone interested in natural phenomena is an emotional experience not soon forgotten. The visitor is soon aware that he is seeing evidence of Nature's complex relationships that have been operating essentially unchanged since the Pleistocene. . . . The value of having such a large natural marsh as Malheur Lake in the National Wildlife Refuge System will increase with each year that passes.[63]

In the 1960s, Malheur Lake was no longer as natural as Duebbert implied. Although he questioned the human attempt to control water systems completely, in 1969 few people in the refuge system agreed with him—certainly not John Scharff. But twenty years later, by the floods of the 1980s, refuge philosophy had shifted toward Duebbert's position, one that questioned the value of such regulation and control.

The 1987 army report acknowledged that project disadvantages might include some damages to the Malheur River trout fisheries but failed to mention other environmental costs and blandly assured the reader that "the plan incorporates measures to mitigate adverse environmental effects."[64] Yet, as refuge staff argued, the environmental effects were far too uncertain and profound to be mitigated by a few miles of riprap to control erosion, or a fish screen to keep carp out of the river. What the canal system to save Malheur Lake from its own variability represented was an attempt not only to control a few floods but to reshape nature in profound ways by changing a Great Basin watershed into a Pacific Ocean watershed.

Malheur Lake is now part of a Great Basin watershed, closed from the sea. This seems part of its essential nature: it is salty, stinky, filled with boron and arsenic and salts and fishes that have been trapped in their closed basins for many millennia. What may be most fundamental to this basin, however, is not its Great Basinness but its variability. At times in the past, Malheur Lake was part of the Great Basin hydrographic system; at other times, however, it was part of the Columbia Basin. The lake once had an outlet to Malheur River, and Virginia Valley (site of the proposed canal) had been the old outlet of the lake to the river. But this was a very long time ago. About thirty-two thousand years ago, rhyodacite flows closed the outlet, forming a dam that backed up an enormous lake in Harney Valley, predecessor of Malheur Lake. But during eras of high precipitation, meltwaters could have flooded the lake so high that it overflowed into Malheur River.[65] Archeological remains show that three fish species now present in the Harney Basin were originally from the Columbia River system— chiselmouth and coarse-scale suckers, and northern squawfish—suggesting that the two systems have probably remained connected since the initial separation.[66] Some scientists feel the Harney Basin may have been linked to the Snake River as recently as four thousand to five thousand years ago, whereas others estimate the most recent separation to have occurred fifteen thousand to seventeen thousand years ago.[67]

Part of what is so striking about the structural solution proposed by the Army Corps is its willingness to reshape nature so dramatically. The differences between a Great Basin watershed and a Columbia Basin watershed seem profound. But is this true in the Malheur Lake Basin, where lake levels have risen high enough in the past to shift the lake from one system to another? Perhaps it is our classifications that are unnatural, not the changes the lake might go through with a canal. Nonetheless, intentionally shifting the lake from one system to another would lead to a cascading set of environmental complications: What would the alkaline waters do to the threatened and endangered salmon runs in the Snake and Columbia Rivers? What would carp do to the native fish communities in the Malheur River? Would they eradicate the native trout fisheries? Would the alkaline, arsenic-laden water from the lakes destroy something essential about the Malheur River, a river already profoundly altered by irrigation?

The Army Corps never constructed this canal system, but not because of environmental consequences. Construction costs had risen too high to justify the new benefit-cost ratio, so economists advised against the project. When floods returned in the late 1990s, so too did murmurs of the old proposal to drain the lake into the Snake River system. But few people took that seriously, because sensitive native species had, in the intervening decade, become much bigger fish to fry.

NATIVE FISH CONFLICTS

In reading decades of Army Corps of Engineers reports about the Malheur Lake Basin, one finds hardly a word about native fish. In 1985 and 1987 the Army Corps did briefly mention the Malheur River trout fishery, but planners dismissed that as being of little importance. For many decades, Malheur Refuge managers as well largely ignored native fish. Waterbirds were the centerpiece of the refuge—in fact, the refuge's name had long been the Malheur Migratory Wildlife Refuge. Even when John Scharff and others paid some attention to fish, it was in the context of killing carp or making people happy by providing them with non-native rainbow trout for sport fishing.

Although carp dominated refuge policies for decades, staff paid attention to them not because they were interested in fish but because carp affected habitat for birds, the important creatures. Eventually, carp did so much damage to Malheur Lake that the refuge staff essentially gave up on the lake as habitat and turned their major attentions to the Blitzen River.[68] Ironically, although non-native fish reduced the value of habitat for ducks, they made it possible for the staff to focus on other elements of biodiversity, such as the few native fish that remained in spite of repeated carp-poisoning projects.

Since the early 1990s, after decades of neglect by managers, a native fish—the redband trout—surprised everyone by emerging as the key species driving restoration and riparian policies in the region. In 1997, after years of drought that had followed the flooding of the 1980s, redband trout in the Great Basin appeared to be in severe trouble. Surveys commissioned by the U.S. Department of the Interior (USDI) and the U.S. Department of Agriculture (USDA) indicated that Great Basin redband trout were extinct in over 72 percent of their historic range, and strong populations remained in only 10 percent of their historic habitat.[69] As one federal fisheries scientist stated, "The status of redbands outside the range of the steelhead, which includes Catlow Valley and other desert basins, is particularly bleak. Redbands in 94 percent of those watersheds are classified as depressed."[70] Although redband trout populations recovered significantly in the late 1990s, they became a lightning rod for concern among environmentalists that riparian areas in the Great Basin had been too long neglected and for concern among ranchers that riparian habitats could derail their visions of transforming the basin.

EVOLUTION

Fisheries biologists now view Great Basin redband trout as an indicator species for the health of desert aquatic and riparian habitats. These fish evolved to survive in high desert environments characterized by unpredictable and intermittent water flows, extreme temperatures, droughts, frequent disturbances, and the harshness of a changing desert environment. In response to such stresses, redband evolved special traits of resilience that have allowed them to survive in conditions that would kill other trout. For example, after drought, redband trout can expand their population sizes more rapidly than other closely related trout species, enabling them to rebound from disturbances and to serve as candidates for restoration as well. But their resilience is not unlimited: it evolved to rely upon certain elements of habitat connectivity, and when that connectivity was broken by human actions, redband resilience seemed to collapse.[71]

Great Basin redband trout are a subset of the species *Oncorhynchus mykiss*, commonly known as steelhead or rainbow trout. Throughout its range in western North America, *O. mykiss* occurs in two separate evolutionary groups: the coastal rainbow trout, west of the Cascade Crest (including the anadromous steelhead that spends part of its life cycle in the ocean); and the redband trout, with life cycles centered east of the Cascade Crest.

Great Basin redband trout are unique among redband trout in occupying closed desert basins that physically isolate them from all other forms of *O. mykiss*. They occur in six areas in eastern Oregon, all lacking access to the ocean: the Catlow, Fort Rock, Harney, Goose Lake, and Chewaucan Basins.[72] Genetic comparisons show that redband trout are genetically distinct from coastal rainbow trout in many ways: they have fewer chromosomes, for example, and differ considerably in allozyme analysis. In turn, Great Basin redbands are genetically distinct from other redband populations, having evolved in isolation over many thousands of years in closed basins.[73] Specialists believe that the Great Basin redband populations represent a group of *O. mykiss* that diverged early in the evolution of the species and has been present in the Great Basin for at least thirty-two thousand years, and possibly more than a million years.[74] Because

Great Basin forms of redbands split from *O. mykiss* early in the evolution of the taxon, they represent what conservation biologists call a significant evolutionary legacy.[75]

Great Basin redbands live in the remnants of great Pleistocene lakes. During the Pleistocene, Great Basin redband thrived in huge alkaline lakes that were interconnected by streams and marshes, allowing fish to move among them. At times, droughts caused the desiccation of various lakes, and redband trout responded by migrating up into streams, where riparian habitat and pools provided refugia. When the lakes refilled, redbands left their refugia and returned to the lakes; this life-history strategy allowed for great resilience to environmental change.[76] But at the end of the Pleistocene, between fifteen thousand and ten thousand years ago, lake levels dropped, breaking aquatic linkages between basins, leaving the redband populations isolated in six distinct populations.[77] If any of these separate populations now go extinct, other redbands could not naturally recolonize the affected areas(unless the lake levels rise again to the levels of the late Pleistocene, something conservation biologists and federal lawyers don't like to count upon).

Because Great Basin redbands have been isolated from rivers that connect to the ocean, they have evolved life histories that are separate from that of anadromous steelheads, their close relatives. Where connections among Great Basin lakes and streams have been broken for thousands of years, some redband populations have evolved a *fluvial* life-history strategy, meaning that they exist only in streams and never migrate to lakes. These fluvial populations spend their entire lives in flowing waters, migrating within the stream system. They usually spawn in the headwaters of their natal streams in the spring (unlike the anadromous redbands in the Columbia River, which usually spawn in the summer). In the fall, fluvial redbands migrate to over-wintering areas within their streams. They eat mostly streamside and benthic macroinvertebrates (not fish).[78]

Where connections between lakes and streams are still intact, populations of Great Basin redbands are *adfluvial,* meaning that they spend part of their lives in streams and part in lakes. Born in the headwater streams, they live there for a year or two, eventually migrating downstream as juveniles into the marshy Great Basin lakes. They spend most of their lives in the huge marshes, feeding on other fish, before reaching sexual maturity at three to five years, depending on environmental conditions and perhaps genetics. Once they reach maturity, they migrate back to their native waters, toward their ancestral spawning grounds. Spawning occurs in gravel substrates, usually in the spring.[79] Perhaps because of the productivity of marsh resources, adfluvial redbands are larger and much more fecund than fluvial forms. Within certain river systems (such as the Blitzen), both fluvial and adfluvial morphs coexist.

Great Basin redbands require healthy instream and riparian habitat for their survival, particularly during droughts. Within the stream, redbands need pools for rearing young, resting places, and refuges from floods and droughts. Spawning requires loose gravel substrates (to provide good oxygenation of stream water, needed for egg survival). High water temperatures are lethal, although redbands are more tolerant than other trout. Biologists find more redband trout in streams with more intact riparian habitat: undercut banks, large woody debris, overhanging vegetation, good cover and shade, healthy streamside vegetation that maintains lower water temperatures, and more habitat for terrestrial insects.

We tend to think that water and land are separate elements, but they interrelate in often surprising and ambiguous ways. On a sunny reach of a river or stream, energy from the sun is incorporated into algae and aquatic plants, moving through the food chain into aquatic insects, fish, then mammals. In a shadier stretch, where too little sun reaches the water to stimulate the growth of algae communities, biological energy washes into the water from streamside grasses, shrubs, sticks, trees, leaves, pine cones, bark, and detritus that has fallen into the river; all of this is incorporated into complex webs of life.[80]

Riparian vegetation showers the stream with nutrients, but most animals cannot digest the cellulose. Small streams are full of tiny invertebrates that turn those trapped nutrients into forms accessible to other animals. Stonefly larvae shred leaves into pieces, snails rasp off wood, beetle larvae carve tunnels into logs, and grazers and shredders break down leaves and woods into smaller pieces. Aquatic insect larvae digest little of the fiber; they poop most of it out into the water, forming an organic soup of shredded leaves and wood, insect droppings, dead insects, and other debris.[81]

For aquatic vertebrates such as redband trout, riparian vegetation provides critical habitat and food. Some of the food comes indirectly, through the invertebrates made possible by riparian vegetation. But much of the food comes directly from the riparian zone. For example, one study showed that half of the food that aquatic vertebrates ate came not from the water but from the land. Where did fish get this terrestrial food, if they were staying in the water where they belonged? The food fell from overhanging shrubs and crawled along the dead and down wood that litters many creeks.[82]

The ability of streams to support diverse and thriving fish communities depends not just on food. Other habitat features are often equally important: pools for rearing juveniles, riffles for feeding, cover for winter survival. Productivity is strongly linked to the geomorphic complexity of a stream: the more side channels, backwater, eddies, pools, riffles, and undercut banks, the better the habitat for many fish.[83] For example, during floods, these channel forms offer critical refugia—places where fish can find shelter from disturbances that would otherwise wipe out their populations. In turn, channel morphology depends not just on geology but also on riparian vegetation and riparian vertebrates such as beaver.

HUMAN EFFECTS

Although native desert trout can withstand brief periods of high temperatures, they can survive only if refugia are present: overhanging banks, pockets of cooler water, subsurface seeps or springs, thick stands of riparian willows. Without these, high temperatures can eliminate redband trout from a stream.[84] Biologists found that on Forest Service land, redband populations were much more likely to be healthy in watersheds with little active forest management or road building, both of which can sever connections between refugia and other instream habitat. Where logging and road building were common, redbands were sparse.[85]

Whatever human activities destroy the connections between water and land will also destroy redband populations. Livestock grazing, because it breaks these connections so profoundly, has been one of the critical factors leading to the decline of redband trout. Livestock degrade trout habitat by trampling and over-grazing, causing erosion and destabilization of stream banks. Shallow, wide streams that result from grazing are poor habitat for juvenile trout rearing or for spawning. Grazing destroys undercut banks and increases sedimentation, which ruins gravel beds needed for spawning; benthic macroinvertebrates are reduced; and grazing can lead to higher stream temperatures, lethal to trout when refugia are gone. For example, 1996 stream surveys by the Bureau of Land Management showed that the three streams in the Catlow Basin rarely met the temperature standards needed for good trout habitat.[86] Researchers have found that where stream and riparian conditions have been impacted by heavy grazing, redband populations are extremely low or extinct.[87]

Irrigation can also prove deadly to redband trout, for, like grazing, it breaks riparian connections. Irrigation dewaters streams, and the isolated pools that remain are unconnected to one another, disrupting stream continuity and trapping fish. Temperatures in these pools rise to lethal levels rapidly. Even when irrigation withdrawals are not pronounced enough to dry up streams, irrigation lowers the flow of streams, shrinking pools, warming the water, and starving riparian vegetation of the water needed to survive. After use on crops, irrigation waters are often returned to streams, but those return flows are hot and contaminated with agricultural chemical and cow manure—not exactly favored redband trout water. Irrigation also leads to loss of fish in unscreened diversions, blockage of migration corridors, and simple destruction of channel morphology. As fragmentation from irrigation decreases the size, complexity, and connectivity of redband habitat, the risk of extinction increases. For example, in a study of a species that needs similar habitat, the Lahontan cutthroat trout, only 32 percent of fragmented stream systems contained the fish, compared to 89 percent of interconnected stream basins.[88]

Fragmentation is especially harmful to Great Basin redbands, for they rely on migration to survive in a variable environment. Migratory corridors allow survivors to disperse following a bad disturbance, such as drought. If habitat has been fragmented, redband cannot migrate away from a desiccating lake bed and thus become stranded. Fragmentation may also subject fish to genetic bottlenecks, thus reducing genetic diversity and resulting in genetic drift. This may happen very quickly: one study found that redband trout in habitat fragmented by dams have significantly less genetic diversity than redbands in unfragmented areas. Fragmentation has had a particularly problematic effect on the adfluvial form of Great Basin redbands because connectivity among marshes, lakes, and streams has been so badly broken. This does not bode well for redbands, since the adfluvial form shows higher rates of growth and reproduction and may be important to the long-term viability of the Great Basin redbands.[89]

What does all this mean for the redband trout in the Malheur Lake Basin? In the Catlow Basin, on the west side of Steens Mountain, three streams still support redband trout populations: Home, Threemile, and Skull Creeks. All three flow into—or once flowed into—Catlow Marsh, which was once Catlow Lake, the largest of the Pleistocene lakes. Before white settlers came, the streams flowed unobstructed into the marsh. During years of high precipitation and high runoff, the trout from individual stream headwaters could reach the marsh, mix, and ascend any of the streams. During low precipitation periods, the streams probably did not reach the marsh, and populations were temporarily isolated, perhaps even going extinct, to be recolonized during wet cycles. But with agricultural, grazing, and irrigation developments, the links among marsh and streams have been fragmented. Fish in the streams no longer can reach the marsh, and so remaining populations have become disjunct; they can no longer intermingle.[90]

The Blitzen River long supported an adfluvial population of redband that migrated to and from Malheur Lake, thriving on the abundant marsh resources.[91] In 1983 fisheries biologists interviewed elderly residents in the Harney Basin who remembered that redband trout were abundant until the 1940s. Marcus Haines, for example, grew up in the Blitzen Valley and worked on a cattle ranch there in 1914 and 1915. He remembered catching twenty- to thirty-inch trout on the Blitzen River each spring, near the tiny community of Frenchglen. Haines told of ranchers dynamiting trout at dams where fish congregated, four miles south of the refuge headquarters. John Scharff recalled that during 1940 and 1942 trout populations in Malheur Lake were extremely high; each spring, he saw many trying to swim up over the dams in the Blitzen. Both Haines and Scharff remembered trout at their most abundant during the early 1940s, and both believed that by the early 1950s few trout were seen below the irrigation dams.[92]

Although adfluvial redbands still persist in the Blitzen River, irrigation withdrawals, weirs, channelization, and draining of wetlands have blocked fish in their attempts to migrate to and from Malheur and Harney Lakes and among surrounding streams since World War II. Carp in Malheur Lake have caused habitat damage so extensive that native trout might never again thrive there.

On Malheur Refuge historic wildlife management practices that focused on waterbirds often harmed redband trout. The water-control system of canals, dams, and irrigation diversions trapped redbands in ditches that slowly dried out, killing the fish. In recent years, refuge staff have worked to restore redband trout through extensive habitat improvements and provision of fish passage and screening. In 1992 managers halted the hatchery program that introduced nonnative trout into the Blitzen River, and all other hatchery programs in the basin are now restricted to reservoirs that do not contain native redbands.[93] Staff are "increasing instream habitat complexity through the addition of rock structures . . . and improving riparian habitat . . . by re-sloping incised banks and then re-establish[ing] riparian vegetation (sedges, willows, rose, etc.)." Other projects under way include restoring connectivity on the Blitzen River, Mud Creek, and Bridge Creek; protecting trout in East Canal; constructing fish passage at dams; and improving water quality.[94]

Elsewhere in the Great Basin, restoration of redband habitat focused on eliminating grazing in riparian zones. Such efforts began in the early 1990s at nearby Hart Mountain National Antelope Refuge (see map 1). Before the Hart Mountain refuge was established in 1936, fifty thousand sheep and ten thousand head of cattle spent entire summers on the mountain, grazing riparian areas to dust. After the establishment of the refuge, biologists began to express concerns about possible impacts of grazing on antelope, motivating slow reductions in cattle and sheep numbers. By the beginning of the drought in 1990, thirteen thousand head of livestock were still grazing the refuge each year. Upland grasses had begun to recover with grazing reductions, but riparian areas remained in poor condition well into the early 1990s because the cattle left on the refuge congregated in the creekbeds.

Severe drought mixed with growing concerns about riparian conditions helped convince managers in 1993 to suspend grazing for five years at Hart Mountain. Local ranchers reacted with fury. Hundreds of public comments claimed that grazing could be compatible with wildlife in a multiple-use management scheme. The refuge staff did not dispute this point, instead countering that, as custodians of a federal refuge, multiple use was not their mandate; the refuge had only one mandate, and that was to maintain the best possible wildlife habitat.[95] Because all available evidence showed that grazing slowed the recovery of damaged riparian areas, no grazing would be allowed until these areas had recovered. Properly managed grazing might not degrade a healthy riparian area, the staff conceded, but since grazing did not accelerate restoration, there was no legal mandate to force grazing to continue.

Death threats failed to budge the refuge staff's position, which was upheld in court. Although Hart Mountain Refuge is not huge, this decision had impacts far beyond its borders. First, it brought riparian areas into the focus of national attention—within the refuge system as well as within environmental and ranching communities. Second, it brought redband trout to the center of the conflict, when refuge staff argued they were required to manage not only for antelope but for all wildlife, including native trout, on the refuge. Third, the decision gave other nearby refuges support in their struggles with local ranchers over riparian conditions. And finally, it led to rapid improvement in redband trout habitat and a rebound in redband populations, encouraging riparian restoration projects elsewhere.[96]

ENVIRONMENTALISTS

The Oregon Natural Desert Association (ONDA), a radical environmental group focused on preserving high desert ecosystems, was little known outside eastern Oregon in the early 1990s. For years, its staff had been trying to draw attention to the problems of grazing in the desert, but change was slow, for few people outside the region knew or cared what a riparian area was, and few wanted to challenge the myth of the cowboy on the range. But in the mid-1990s, in the wake of Hart Mountain's decision to remove cattle and in the face of growing regional concern over salmonid habitat, ONDA found a wider audience.

The group rose to prominence when it placed the Clean Stream Initiative on the Oregon ballot in 1996. The initiative required that "the waters of the State of Oregon shall be protected from water pollution caused by livestock."[97] For the first five years, the act would have applied only to public lands that supplied drinking water or trout, salmon, or steelhead habitat; after ten years, it would have applied to all of Oregon's waters.

On the face of it, keeping streams free of cattle manure seemed like a simple requirement, given that Oregon waters belong to the public, not to individual landowners. But because the proposed legislation would have affected not just streams but also their riparian areas, ranchers reacted with fury, eventually leading to the defeat of the initiative.

Although many Oregonians were concerned about trout, the perceived threat to private property rights proved more powerful. Although Oregon's waters are public, its riparian areas are on private land. Landowners loathed the idea of others telling them what they could do with their private land, and the anti-initiative campaign drew attention to the issue of public access onto private property. The act would have allowed any citizen to bring a civil action against "any person, including the State of Oregon, alleged to be in violation" of keeping livestock out of waterways. This became the major bogeyman for the public: opponents blanketed the airwaves with scenarios in which a nice family was sitting down to eat, and radical environmentalists came boldly onto their private land to snoop on them and take them to court. Because riparian areas confuse the clear boundaries between water and land and public and private, efforts to protect them with the Clean Stream Initiative failed.[98] But the fact that the initiative got on the ballot at all, and that intense publicity was needed to defeat it, meant that riparian areas, grazing, and the effects of cattle on redband trout habitat became headline news. Riparian areas were no longer the invisible sacrifice areas, and redband trout were no longer the invisible vanishing fish.

While campaigns raged over the Clean Stream Initiative, environmentalists were also using another arena to protect redband trout habitat: the courts. Legal battles focused on the Donner und Blitzen River, particularly on lands managed by the Bureau of Land Management on the west slope of Steens Mountain, which supported populations of wild redband trout. In 1988 Congress had designated much of the Blitzen a wild and scenic river. Yet the next year, in the 1989 Andrews Resource Management Plan, the Bureau of Land Management recommended that AUMs for cattle in riparian areas along the Blitzen River should increase, rather than decrease. Environmentalists had been fighting the bureau over grazing in the Wild and Scenic River corridor ever since, but the bureau had refused to remove cattle from riparian zones within the corridor.

In 1995 the Bureau of Land Management published a new plan for the land it managed along the Blitzen River (technically, the South Steens Allotment). Although acknowledging that riparian quality and redband trout habitat were in poor condition along much of the watershed, the bureau refused to prepare an environmental impact statement for its grazing management there. In response, the Oregon Natural Desert Association took the federal agency to court.[99]

Ironically, it was the Bureau of Land Management's own staff biologists who had shown that water quality was extremely poor along much of the Blitzen drainage, so poor that it endangered what remained of the redband trout populations. The Bureau of Land Management's 1995 "Plan for the Blitzen River" reported numerous examples of severely degraded riparian habitat, concluding that "significant portions of many streams have degraded fish habitat as exhibited by a lack of woody riparian cover; unprotected, often unstable, streambanks; [and] poor instream structure due to sedimentation." After noting that poor habitat was linked to BLM grazing practices, the report urged, "We must begin to implement improved management immediately."

Even with their own research showing that fish habitat quality was poor and that grazing practices in riparian areas on BLM land were largely the cause, the managers still decided that an "Environmental Impact Statement is unnecessary and will not be prepared."[100] Their reasoning was based on historical interpretations. The bureau argued that if riparian condition was improving, no matter how slowly, that was good enough. And since riparian condition had been so poor during the early years of the twentieth century, no matter how bad it was in the 1990s, that was still an improvement over the past. No matter how slow the improvement was, and no matter how much faster the improvement would be with grazing changes, the bureau still interpreted any improvement over the old days as legally sufficient.[101]

Rather than reducing AUMs, fencing off riparian zones, or eliminating cattle from redband trout habitat, the Bureau of Land Management decided instead to simply allow ranchers to try using riders to herd cattle out of damaged streams. Surprisingly, even the bureau had little faith in the ability of riparian areas to heal under these conditions. As the plan noted, improvement would

occur only if herding with minimal fences proves to be a viable alternative to additional pasture fencing. . . . Herding success can vary within different stream reaches. . . . Success of herding is dependent on knowing the habits of cattle grazing during different seasons, cattle preferences for plants at different elevations, and the skill and commitment of the riders.[102]

The Oregon Natural Desert Association took the Bureau of Land Management to court for this decision, arguing that the agency was violating the terms of the Wild and Scenic Rivers Act, which required management to "protect and enhance river values," and also violating the National Environmental Policy Act (NEPA) by failing to prepare an environmental impact statement for the action. In 1997 federal judge Ancer Haggerty ruled against the bureau, finding that because cattle were indeed damaging redband fish along the Blitzen River, grazing needed to be reduced. Haggerty ruled as well that the agency had violated both the NEPA and the Wild and Scenic Rivers Act by not working to "protect and enhance" river values. He disagreed with the agency's use of history, arguing that the goal of an "improving trend in riparian condition" was simply not good enough.

The judge ordered the Bureau of Land Management to meet with the plain-tiffs to reach some agreement on how many cattle could graze on BLM land along the Blitzen River in the next year, threatening them with a court-ordered graz-ing plan if they failed to agree. Most significantly, this decision brought old ene-mies to the negotiating table, forcing the bureau, private ranchers, and the Oregon Natural Desert Association to construct a conservation agreement that would restore redband trout populations.[103]

EFFORTS TO RESTORE REDBAND TROUT

Although the Bureau of Land Management had joined in the creation of a con-servation agreement, it continued to insist that cattle had not contributed much to the damage and that any grazing damage was due to overgrazing in the dis-tant past, not to current practices.[104] Frustrated with the slow pace of coopera-tive efforts, in September 1997 the Oregon Natural Desert Association, Oregon Trout, the Native Fish Society, and the Oregon council of Trout Unlimited peti-tioned the Fish and Wildlife Service to list the Great Basin redband trout as threat-ened or endangered under the Endangered Species Act. The petitioners argued that redband populations were continuing to decline and that existing regula-tions were inadequate to protect the trout from extinction.

On March 13, 2000, the Fish and Wildlife Service announced that listing the Great Basin redband trout as endangered under the Endangered Species Act "was not warranted at this time." The agency did not dispute that the fish had been eradicated over most of their historic range or that human activities had harmed them. But because stream surveys in 1999 found that redband populations were increasing after the drought, the agency argued that the species was recovering and therefore not in need of protection: "Because redband trout populations in all basins have rebounded, the effects of any potential threats to the Great Basin redband trout and the likelihood of extinction of the species is substantially reduced." Redband habitat had been degraded, the agency concluded, but since there was enough good habitat to support an increasing population after the drought, the agency argued that habitat was likely good enough to keep the fish from going extinct.[105] In biological terms, this seemed to many an odd decision. To merely note that redband populations were increasing after the drought might not necessarily mean much for the long term viability of the redband, since no matter how poor the habitat is, after a drought populations are likely to increase. If all one needs to prove that a species is not threatened is to show an increase in numbers after an extensive natural perturbation, then it might prove impos-sible to list any species that lives in a variable environment.

The major reason for rejecting the petition may have been political rather than biological: the Fish and Wildlife Service wanted to encourage private and cooperative efforts at restoration. The decision cited the conservation agreements on the Roaring Springs Ranch, in which the Bureau of Land Management had been ordered by Judge Haggerty to come up with a new management plan to protect the habitat. The conservation agreement had listed seventy-four actions that would be undertaken by the ranch and the Bureau of Land Management to protect fish habitat, and the Fish and Wildlife Service found that most of these had been completed or initiated, concluding that "based on the cooperation between private, State and Federal parties, most if not all of the current threats to redband trout in the Catlow Basin are being or will be addressed."[106]

"Cooperative efforts involving all parties are excellent avenues for restoring habitat and species," the Service wrote in its decision, making clear that it feared that a decision to list the Great Basin redband trout might have threatened the future of other cooperative efforts.[107]

Conservation groups responded to the decision with disappointment. The conservation director of Oregon Trout, Jim Myron, stated that although redband trout had indeed rebounded after the end of the drought, "all the adverse habitat conditions affecting these fish that were presented in the petition are still valid and will likely worsen in the future." John Gelbard of the University of California, Davis, the scientist who prepared the petition, pointed out that because the Fish and Wildlife Service's numbers on redband recovery came from surveys made during wet years, they "may reflect the high end of a population cycle in the basin. The big danger would be during drought years."[108]

The Oregon Department of Fish and Wildlife, on whose surveys the Fish and Wildlife Service had based the decision not to list, cautioned that its own data were not enough to assure complacency. Wayne Bowers, the fisheries biologist in Hines who did much of the field sampling, warned, "There's going to be dry years to come and we need to protect healthy stream habitat for the fish to have a safe refuge." Chip Dale, the regional director of the Oregon Department of Fish and Wildlife, agreed, saying, "With the end of drought and good water year in 1999, redband trout numbers were sufficient to stave off a listing. But efforts need to be ongoing to protect habitat for the fish, so it can survive the next drought." Like the Fish and Wildlife Service, Chip Dale focused on the success of cooperative agreements, saying they "were the formula for success in this effort, and I think they point the way to the future."[109]

Throughout the region, government agencies interpreted the Fish and Wildlife Service's decision not to list redband as a success for the cooperative agreements designed to use voluntary efforts to avoid listing of a species. Roaring Spring's decision to enter cooperative agreements was cited as proof that ranching and trout could coexist. Voluntary agreements, many said, were the future of conservation, for they could avoid head-on environmental collisions that might threaten the future of the Endangered Species Act.[110] This particular voluntary agreement led to real changes on the ground, for the Roaring Springs Ranch has done excellent restoration work. Nevertheless, one sobering lesson from this history cannot be ignored: without the court battles, few of these cooperative agreements might have gotten anywhere. Cooperative agreements in the absence of legal or political power mean little.

Attempts to restore riparian habitat were dramatically altered when Secretary of the Interior Bruce Babbitt made a surprise visit to Steens Mountain in August 1999. Appalling ranchers and delighting environmentalists, Babbitt announced that he intended to "protect the mountain as a federal treasure"—meaning a national monument, national park, or national conservation area.[111] Babbitt pledged, however, to preserve ranching on Steens Mountain—a promise that met with skepticism from the thirty-five ranchers (with their eighteen thousand head of cattle) who used the mountain.

Babbitt assured ranchers that the Clinton administration would not simply declare Steens Mountain a national monument, as had happened with Grand Staircase–Escalante National Monument in southern Utah in 1996. Instead, Babbitt made both a promise and a threat, telling the fifteen-member Southeast Oregon Resource Advisory Council—a group made up of ranchers, business people, an environmentalist, and tribal representatives—that they had two months to come up with a plan of their own to protect the mountain, or else he would declare it a national monument.[112] Rather than suggesting creation of a national park or monument, either of which would have been administered by the Department of Interior under a strict set of general regulations, Babbitt proposed a national conservation area administered by the Bureau of Land Management, which would allow for much more flexibility.

Established early in Babbitt's tenure with the hope of increasing local participation in BLM decisions, resource advisory councils included ranchers, outfitters, environmental groups, tribal representatives, and other community members. These councils had unfortunately accomplished little, for different stakeholders were set into their positions and had little motivation to negotiate.

For example, one member of the Southeast Oregon Resource Advisory Council, Fred Otley, had originally opposed any additional protection for the area. Otley is a fourth-generation rancher on Steens Mountain in the Kiger Creek and Kiger Gorge area. He is the coordinator of Friends of Steens Mountain, a group of local citizens and ranchers, and is a "private landowner liaison" to the council. The Otleys had managed their private lands so well that creeks on their holdings had the healthiest redband populations of any sites monitored by the state. Otley was frustrated that the private ranchers who had restored and protected riparian habitat might end up losing the most if Steens Mountain got additional federal protection.[113]

Another member of the council was Stacy Davies, manager of the Roaring Springs Ranch, the largest ranch on the mountain, with 146,000 acres of private land and 250,000 acres of BLM leases. Like Otley, Davies was initially opposed to federal protection of the mountain. In an interview with *High Country News,* Davies said, "Basically, our ranch isn't for sale. . . . If you make it a federal area, it would destroy the area that you're trying to save. . . . I think we may be a year away from losing the whole thing."[114]

Many environmentalists were equally unwilling to negotiate. For example, Bill Marlett, the executive director of the Oregon Natural Desert Association, initially opposed any compromise that would allow any grazing on the mountain: "I told [Babbitt] point-blank that we want a date-certain phaseout of grazing. We just don't think that cattle grazing is compatible with the desert environment. . . . We do not want a national conservation area. Period."[115]

Not surprisingly, the Southeast Oregon Resource Advisory Council was initially unable to find a consensus on the mountain's future. In October 1999 the council submitted a report opposing a new designation for Steens Mountain, arguing that most local landowners did not want more protection for the region. Babbitt, however, told them to go back to the drawing board and work on the outlines of a plan with the governor and with Republican Greg Walden, the U.S. representative for Harney County and much of southeastern Oregon. Babbitt tried to convince the council that a plan drafted by Oregonians would be better than "letting outsiders decide what's right for Steens Mountain."[116]

Fred Otley and his neighbors decided drastic measures were in order to save their family ranches. With Otley serving as their chief negotiator, they proposed that one hundred thousand acres in the highlands of Steens Mountain become a "no-grazing" wilderness. In exchange, they wanted a voice in deciding how other public lands on the mountain would be managed, plus assurance that they would be left to run their private operations as they saw fit. Finally, just before the end of the congressional session in July 2000, the Oregon delegation introduced the product of the resource advisory council's efforts: HR 4828, the Steens Mountain Cooperative Management and Protection Area Act, which designates 143,000 acres of the mountain as wilderness, retiring grazing permits from 100,648 acres mostly along riparian areas in the Blitzen and in the Catlow Basin. Many, but not all, streams and rivers receive protection from grazing. Ranching is allowed on much of the rest of the nine hundred thousand acres, and ranchers are eligible for thirty-year grazing permits (compared to the current ten-year BLM permits) when those permits are coupled with long-term conservation easements on private land, or nondevelopment agreements.[117] Local ranchers have a key voice in determining management, under the condition that environmental laws are not violated. In return, the legislation prohibits subdivision of the mountain for resorts or vacation homes. In early October Congress voted to pass the bill, and on October 30, 2000, President Clinton signed it.

Perhaps what was most striking about this turn of events was that people from such different perspectives managed to compromise. What changed in the political landscape to make this possible? Bruce Babbitt's threat of national monument designation was certainly a powerful motivator, as was the threat of additional litigation from environmental groups. Yet threats were not the only thing that mattered. Groups with radically different perspectives had managed to find some common ground: their shared distaste for development and subdivision by Californians. Otley was key in negotiating this compromise, not because he suddenly turned his back on ranching, but because he was able to envision a future in which his identity was not inalterably opposed to environmentalists.

Riparian areas, invisible just a few years before, had become enough of an issue to determine the terms of the compromise. Managing grazing to restore redband trout habitat and removing livestock from riparian zones to allow for recovery: these were no longer the wild-eyed suggestions of radical environmentalists but consensus among the most conservative of ranchers, county commissioners, and Bureau of Land Management range conservationists.

Riparian areas were not only a new category of analysis; they were the category with the greatest power to drive policy and law. The invisible had become profoundly visible. But another lesson from this is equally important: before the threat of legal action by environmentalists and administrative action by the federal government, riparian restoration efforts had not gotten anywhere. In a world where economics constrain what ranchers can do, ideas and good intentions alone would never have protected redband trout. Politics and the law were equally important.

6 / Pragmatic Adaptive Management

n the Malheur Lake Basin, Paiute hunters and gatherers, white ranchers, irrigation developers, speculators, engineers, and wildlife biologists all have competed for control of the uncertain boundaries between water and land. The cultures of each defined boundaries differently. Each group told a different story about its own relationship to those boundaries and about its own right to transform them. These stories had, and still have, tremendous power in helping to shape Malheur.

In 1878 the Paiute told a story that went something like this:

> We were living here for thousands of years, until the whites came like a roaring lion. They made the grass, the animals, the water, and the willows vanish, and then they tried to kill us.

Now the story the Paiute tell is different, for it is grounded in renewed economic and political power. The Burns Paiute have constructed a casino, have legal standing as a tribe, have been regaining their water rights on their former reservation and have become a key stakeholder in the resource advisory council's effort to shape policy on federal lands.

In 1878 the ranchers told a story very different from that of the Paiute. In their eyes, the Paiute were hardly human:

> We came to an empty land where the water and the grass went to waste. We brought in cattle, we took all the risks, and we built an empire here, a place of good homes and good cattle. We moved water across the land, making the dry lands wetter and the wet lands drier, and that labor gave us rights to that water and that land.

Like that of the Paiute, the story ranchers now tell has changed, for the balance of power has shifted in the basin. Instead of determining basin policy, ranchers are now one of several stakeholders. They still have a great deal of power, but they do not have all of it.

In the 1890s and early 1900s, homesteaders told a story in which both the Paiute and the ranchers possessed little right to the land:

> We came to a place that was still fundamentally empty, with just some cows and cattle barons wasting water and wasting land that could be turned into a prosperous, American place. We made the desert bloom against all odds—against the greedy cattle barons who refused to let us use the water they were wasting, against a government that promised much and delivered nothing, against the weather, against the barren marshes, against the alkaline soils, against nature herself.

The wildlife biologists who worked so fiercely to preserve Malheur told a story in which homesteaders were fools tricked by greedy speculators into a doomed attempt at farming. John Scharff told a heroic story in which conservationists saved the birds:

> We saved them from irrigators and homesteaders and drought and predators and from nature herself. With the help of engineering, we made a brave new world that maximized production of grass, cattle, and waterfowl.

In the years since Scharff retired, management of Malheur riparian areas has become less clumsy but no less manipulative. Now, instead of using bulldozers to channelize the river, the staff is trying to figure out ways to use bulldozers to return the river to its old meanders. Willow are being planted instead of being ripped out, but herbicides still play a role, removing vegetation that might compete with desired native species. The irrigation and water-control system grow ever more elaborate, since without it much of the habitat for rare and endangered birds would be lost. Flood irrigation still waters the meadows, but now it creates hay for bird cover, not just for cattle.

The most profound change in the Blitzen Valley is that refuge staff are no longer trying to fix a single pattern of ponds and meadows and wetlands. Instead, they are trying to manage variability back into the system by alternating which meadows are dry and which are wet. Yet, given the constraints of managing a wildlife refuge with extensive investments in structural improvements, this variability can be allowed only within strict limits. For example, the river may be allowed to meander a little, but not enough to threaten the constructed canals and brood ponds.

Variability in predator and prey population cycles is allowed on many refuges, but only within certain limits. On nearby Hart Mountain, refuge manager Mike Nunn decided to reduce coyote predation on pronghorn antelope through aerial hunts, since the antelope are rare and their predators are not. Many environmentalists are appalled by continued efforts at predator control, seeing it as evidence that federal managers are stuck in old wildlife management paradigms aimed at maximizing single-species production. When environmentalists challenged Hart Mountain's plan in court, the Fish and Wildlife Service retreated, saying the plan would be postponed until they had time "to redefine the purpose of the 275,000 acre refuge and clarify the roles of pronghorn and coyotes."[1] But environmentalists remain unconvinced that refuge managers can ever abandon their faith in maximizing antelope numbers through predator control. Joy Belsky, an ecologist with the Oregon Natural Desert Association, said, "As long as we have Mike Nunn in charge, we are going to have the same outcome." According to the *Portland Oregonian*, Belsky believed Nunn was "obsessively concerned with coyote predation," and she doubted Nunn could be persuaded not to kill coyotes: "We have an historical record now of what Mike Nunn plans to do."[2]

Critics of predator control believe that the effort to kill coyotes in order to save antelope is a case of bad science. On this point, they may be correct, given that the refuge had not completed surveys of antelope numbers, recruitment of young antelope into the population, coyote numbers, or the impact of coyote predation on antelope reproduction relative to the impact of sagebrush expansion. Political pressure may well have influenced the refuge's decision to move forward with a coyote hunt before it had completed all the scientific work needed to support its proposal.[3]

But the objection to predator control goes far deeper than a critique of available science. Critics feel that predator control violates the idea of natural regulation— the hope that predators and prey can find their own balance in the absence of human intervention. Leave coyotes alone, critics demand, and wait and see how the ecosystem repairs itself. Or, as the Predator Project declares, "The Refuge provides a textbook example of what happens to an ecosystem when humans choose to become involved."[4] Human intervention in nature, critics feel, is the core problem. A wildlife refuge should be free of human intervention—in natural places, natural processes should be allowed to function without human management. Joy Belsky says she hopes that "nature is allowed to assert itself" across the Great Basin refuges—which means "allowing wildlife to undergo natural population cycles."[5] But what exactly does it mean to be natural today? Magpies are killing so many cranes—and coyotes may be killing so many antelope—not because such predation is natural but because human interventions have created hybrid landscapes in which magpies and coyotes flourish.

Likewise, some critics of Malheur Refuge policy have recently argued that the water-control system should be dismantled and that natural variability should be allowed to have full sway. But is this possible in a world so dramatically altered by people? Before the advent of extensive water control systems, water levels were so high some years that numerous pools and ponds formed in the valley, perfect for brooding waterfowl. Other years few pools formed, and waterfowl-rearing habitat was minimal. This historic variability existed within an entirely different context, however. Malheur Refuge was once only one of a long string of fertile, vast marshes stretching up and down the Pacific Flyway. Much of the Great Basin was stopover habitat for migrant birds needing to rest and fatten up on their long journeys to the Arctic. If most of the Malheur Lake Basin happened to be dry one year, the birds could stop elsewhere, because the Pacific Flyway consisted of numerous patches of desert, riparian, and wetland habitats.

Now, however, the vast majority of those historic riparian areas and marshes have been lost to agriculture, shopping malls, and highways. Malheur Refuge has become critical habitat in a way it never was before. If natural variability were returned at Malheur, it might be disastrous for entire populations of ducks, sandhill cranes, and shorebirds. Until millions of acres from California to Canada have been restored back to wetland and riparian meadow, allowing natural systems to work entirely without human intervention is as unnatural as trying to gain control over every drop of water and every act of predation.

Refuge managers feel, in other words, that they cannot allow natural systems to be purely natural. Managers try to restore some natural variability, but not enough to threaten the water systems that have been painstakingly constructed. There is nothing ideologically pure about current refuge policy: it is not an attempt to return to pristine natural conditions, nor is it an attempt to gain complete control of nature.

Such a policy infuriates some environmentalists, who see little difference between John Scharff's regime and current refuge attempts to regulate water. But this misses crucial differences. Scharff, unlike current managers, did aim for ideological purity: his ethic was one of control and improvement. He rarely seemed to doubt that humans could and should take complete control of nature. Likewise, some modern environmentalists have an ethic that is equally ideologically pure: the ethic of naturalness. A thing is good when it is natural, bad when it is not. Controlling predators or water is unnatural, so therefore it is bad.

In the world that refuge staff actually have to work in, neither ethic is particularly helpful. Staff struggle along, trying to find some reasonable path between extremes, zigzagging back and forth between management of ducks and cranes and fish. The refuge managers are trying to act pragmatically rather than ideologically. They are not trying to restore the refuge to some past set of pristine ecosystems, but to adapt to change, making things work as best they can while minimizing future complications.

Such pragmatic decisions are the key to adaptive management, which is the messy process of developing a management scheme that incorporates multiple human perspectives while responding to changing scientific understanding of dynamic ecosystems. At its best, adaptive management is a way of paying close attention to what happens when we manage landscapes and then altering practices when old ways no longer produce the desired results (or when the results that people desire change). This, at heart, is simply applying the scientific method to management. Everything managers do is nothing more, and nothing less, than an experiment. Experimentation means approaching the world with an open mind. As a scientist, one is supposed to treat one's own ideas with humility, abandoning hypotheses if results are as expected. This process is never completely open-minded; initial ideas about how the world ought to work shape what one sees. But there is an important ideal here—that of allowing the natural world to shape human ideas, and not just the other way around. In other words, there is a kind of give and take, a willingness to be surprised. The critical step for management, however, comes after the research, when new information must be used to change how one works with the land. Adaptive management does not necessarily mean big government programs. Above all, it means people on the ground being responsive to what the land is telling them, and being responsible for acting on that knowledge. It means a dialogue between people and land; it means people knowing the place they work.

Managers such as John Scharff long hoped that they could engineer the riparian landscapes to produce stable outputs of what people most desired, but the watery landscape proved far too dynamic for this. No matter how many facts managers accumulate and how many theories they test, they will never have the knowledge to manipulate natural systems without causing unanticipated changes. Yet they still have to manipulate the environment, which presents a dilemma: How does one make decisions when one knows one will never be able to fully predict the outcome of those decisions?

Adaptive management at its best is an iterative process that yields new information about ecological and human systems and then uses that information to develop policies that can respond to changing knowledge about a changing world. Although few proponents of adaptive management have been trained in American philosophical thought, framing adaptive management within the context of American pragmatism can help us understand why the process offers promise for resolving environmental conflict in productive ways.

American pragmatism emerged at the beginning of the twentieth century in the works of Charles Peirce, William James, and John Dewey, who argued that there are no innate givens upon which our knowledge is built. As James put it, "Novelty and possibility [are] forever leaking in."[6] Therefore, he argued, a pragmatist turns away from "*a priori* reasons, from fixed principles, closed systems, and pretended absolutes and origins. He turns towards concreteness and adequacy, towards facts." For James, pragmatism meant "the open air and possibilities of nature, as against dogma, artificiality, and pretense of finality in truth."[7]

Skeptical of absolutes, distrustful of ideology, pragmatism tries to be useful—it cares less about ultimate, unknowable distinctions or essences, and more about what can contribute to resolving policy conflicts and forming democratic communities. Three elements of pragmatism are important for adaptive management: concern with evolutionary adaptations in a changing environment; interest in the scientific method as a process for understanding that changing world; and concern with pluralism, multiple voices, and the formation of democratic communities.

James and Dewey developed their philosophical theories in the context of Charles Darwin's recognition that an organism exists in relation to a changing environment.[8] From their interest in evolutionary theory, pragmatists developed a profound sense of what Dewey termed "a world still open, a world still in the making."[9] Evolutionary ideas convinced them that striving for absolutes is counterproductive, since indeterminacy and change are always central features of a "world still in the making." What matters is the interaction between the human organism and its dynamic environment.[10] Dewey and James saw individuals not as "single entities that simply stand alongside one another in the world" but rather as creatures defined by relationships, by "connections, transactions and entanglements." They believed that humans, like other animals, are "embedded at every point in the broader natural sphere." Identities emerge not in isolation, but "in the ongoing negotiations between humans and environments."[11]

Above all, the pragmatists believed that theories of evolutionary change teach us that the environment can never become "a fully settled, predictable thing" and that no individual or group can ever have complete knowledge of it. The natural world always escapes our stories about it. The earth should not be viewed as an instrumental resource for human mastery, since we can never know enough to declare parts of it worthless.[12] James wrote that humans should not "be forward in pronouncing on the meaningfulness of forms of existence other than our own." Rather, we must "tolerate, respect, and indulge those whom we see harmlessly interested and happy in their own ways, however unintelligible these may be to us. Hands off: neither the whole of truth nor the whole of good is revealed to any single observer."[13]

No one, James believed, can ever possess the absolute knowledge of unchanging truth that would allow judgment of the value of other species, for truth is always contingent on history, environment, and context: "The world is full of partial stories that run parallel to one another, beginning and ending at odd times. They mutually interlace and interfere at points, but we cannot unify them completely in our minds."[14] Theorists alone can never provide all the answers, James believed, for their "concepts can only provide a static picture of a world which is essentially dynamic."[15]

Pragmatists argued that in solving problems, we do better not by reasoning from first principles but by applying the scientific method. Ideally, one forms an initial hypothesis based on past experience, and then one tests that hypothesis against other actors in the system and against the available data. In science, as in pragmatism, a theory is true if it "works"—if it can be tested and supported. Experience and experiment are at the core of this process. Knowledge emerges from active experimentation, and that knowledge is contingent on place, time, and experience.

The central question for adaptive management is, How can resource managers encourage such active experimentation, given the many forces—social, political, and economic— that encourage inertia? James argued that we are continually trying to make sense of the world through "the matrix of conceptual constructs . . . that bring order to raw experience." Yet, as James added, "the world we live in is surrounded by a fringe of the unknown . . . that is larger than ourselves and our settled knowledge. It is on this fringe, and in those parts of our knowledge that occasionally become unsettled, that the transformative activity of knowing goes on."[16]

When old ways of knowing no longer work, "we must find a theory that will *work;* and that means something extremely difficult; for our theory must mediate between all previous truths and certain new experiences."[17] New theories cannot arise in isolation; they must make sense of our received cultural legacies and of our new experiences in a changing world.

At Malheur, refuge managers in the 1930s made reasonable engineering decisions in a desperate situation, but by the 1940s they proved slow to respond to indications that their schemes were leading to trouble. When events at the refuge began to spiral out of control, managers didn't question their own basic assumptions but instead tried to hold the system under increasingly rigid control. This, as the pragmatists noted, is hardly unusual: people rarely respond well to experience that contradicts their received ideas. Adaptive management assumes that managers can quickly discard old ways that don't work, but the reality is that few people willingly abandon the stories that give meaning to their lives. As James wrote, "By far the most usual way of handling phenomena so novel that they would make for a serious re-arrangement of our preconceptions is to ignore them altogether, or to abuse those who bear witness to them."[18] This, of course, is what has often happened in federal agencies charged with land management: John Scharff, for example, like many other managers, found himself unable to undergo a "serious re-arrangement of his preconceptions." Adaptive management will work only if people recognize this difficulty and devise processes for encouraging experimentation and change. John Dewey believed that revision of received ideas could happen only when individuals are part of a larger, questioning community; progress requires "the maintenance of social structures that encourage continuous inquiry."[19]

Management techniques at Malheur that began as experiments soon became orthodoxy. In the 1930s the refuge staff developed a set of powerful techniques that made excellent sense in a particular context, given the challenges waterfowl populations faced at the time. But as Scharff gained power, those ideas became increasingly rigid. People found it difficult to challenge the developing orthodoxy until outside events—floods, litigation, and the threat of listing redband trout as an endangered species—forced refuge managers to take new perspectives seriously. Conflict forced people, institutions, and states to incorporate new ideas into their worldview.

For generations, first ranchers and then refuge managers were able to gather enough power so that they did not need to acknowledge viewpoints other than their own. Ranchers moved away from their hold on the basin only when lawsuits, drought, and financial ruin in the 1930s forced changes in the balance of power. Four decades later, refuge managers were reluctant to move away from their own ideologies until environmentalists used litigation against them. A set of escalating conflicts, which began as local issues and then became mediated by national institutions, eventually forced groups in Oregon to embrace a political process in which stakeholders coming from different perspectives had to jostle against one another, argue with one another, and listen to one another in ways that modified their actions and beliefs. Because no one has perfect knowledge of how ecological systems work, this process moved participants toward much better solutions than any one group could have found on its own.

An important claim I have made in this book is that the refuge wouldn't have made such progress if not for litigation and other conflicts. What mattered most about litigation was not that the judges finally learned some essential truth from environmentalists—in fact, many of the ideas promoted by environmental groups during these lawsuits will probably turn out to have been wrong. Rather, litigation forced a variety of stakeholders—with multiple voices, stories, and perspectives—to communicate with one another. This led to new ideas and eventually new conditions for the refuge. Refuge managers ran into trouble when they were permitted to operate with the authority of state power reinforcing their assumptions. Only when political conflict in the 1970s forced managers to allow other stakeholders to have a voice did the refuge managers begin to question some of the bad assumptions that seemed so self-evident when they essentially didn't have to answer to anyone else.

In *Seeing Like a State,* James C. Scott asks why so many schemes to improve nature—schemes that, like Scharff's, were instituted with the best of intentions— have led to disaster. He argues that these programs failed, in part, because they were based in what he calls "high modernist ideology. . . . A self-confidence about scientific and technical progress . . . and the rational design of social order commensurate with the scientific understanding of natural laws." By failing to attend to "the indispensable role of practical knowledge, informal processes, and improvisation in the face of unpredictability," such ideology ignores the practical knowledge and the lived experience of local communities.[20] Big ideas, Scott argues, become dangerous when they become an ideology that freezes itself off from questioning, in the way that Scharff's ideology became isolated from the contingency, uncertainty, and complexity of life as actually lived.

As at Malheur, the failures Scott examines in Europe, Latin America, and Asia were not "cynical grabs for power and wealth" but rather were "animated by a genuine desire to improve the human condition." In his case studies, high-modernist interventions came "cloaked in egalitarian, emancipatory ideas"— something certainly true of reclamation projects in the Malheur Lake Basin. Nevertheless, these interventions were products of hubris, for planners rarely began "from a premise of incomplete knowledge," and they routinely ignored "the radical contingency of the future." As Scott argues, pragmatic approaches to resource management would be more prudent:

> A step-by-step "muddling through" approach would seem to be the only prudent course in a field like erosion management or public policy implementation, where surprises are all but guaranteed. . . . One would want hydrologists and policy managers who had been surprised many times and have had many successes behind them.

Such experience allows one to "interpret early signs that things are going well or poorly"—in other words, to practice adaptive management. In a pragmatic approach to management, Scott argues, managers should take small steps, monitor the results, and adapt their next steps accordingly. They should use an experimental approach recognizing that "we cannot know the consequences of our interventions."[21] Put into practice, an experimental approach would favor reversibility—use of interventions that can be undone easily if necessary.

Ideologies alone are rarely enough to impose order on landscapes and communities: managers also need power.[22] At Malheur, during the earliest years of the refuge, managers such as Ira Gabrielson did not have authoritarian state power, for they had to contend with competing groups' claims to the refuge. Homesteaders, ranchers, and conservationists all required negotiation. John Scharff's silencing of these alternative voices made him more efficient but this ultimately led to trouble. My argument throughout this book has not been that federal land managers are dangerous people who strive too hard to control nature. My argument is pragmatic: orthodoxy is dangerous, for it assumes complete knowledge in a world where such knowledge is simply impossible. Managers become dangerous when they close themselves off from challenges to orthodoxy and when they use the power of the state to enforce their vision on human and ecological communities.

Although state power and narrow scientific expertise can enforce rigid and dangerous orthodoxies, the answer is not to take power away from the scientists or the state and simply give control of resources back to the locals. Much of the recent scholarship in environmental history, mine included, has pointed out numerous problems with the progressive conservation movement and its legacy of elitist expertise, which used state power to control both the natural and social worlds. In a recent book on civic environmentalism, *The Land That Could Be*, William Shutkin notes that environmental historians have shown that "environmentalism has essentially elitist roots" that helped shape it as a "complex system of laws and policies fixated on preserving undeveloped land and resources."[23] Shutkin, like many other current critics of environmentalism's supposedly elitist past, wants to see environmentalism move "beyond regulations and adversarial legal approaches," believing that "top-down initiatives are automatically suspect." Instead of a reliance on regulation, he wants "innovative, dynamic collaborations among stakeholders" in a region. More important, he wants local people to have the power to determine what happens on their land.

Shutkin is certainly no Wise Use antienvironmentalist—he himself is doing innovative, dynamic work with urban communities that were often overlooked by some parts of the modern environmental movement. But he fails to recognize that despite the mistakes made by technocratic, scientific experts, their expertise has been useful and necessary. Even if members of local communities live closely on the land, that doesn't mean scientific expertise has nothing to offer them. Although top-down regulatory approaches have led to some terrible problems, the alternatives aren't necessarily any less problematic.

Locals in the Malheur Lake Basin often say, "I live here, I know this land, I love this land. Who are you to tell me what I can and can't do on my land? You experts think you know more about my land than I do? You couldn't possibly!" But the truth is that although people who live and work on the land do have experience that is critical for good land management, they lack a great deal of knowledge that is equally critical for responding to change. Scientific experts do know more about certain elements of the landscape than the farmers and ranchers who have spent their lives on it. That doesn't mean we should ignore local knowledge, but neither does it mean that local knowledge is always sustainable.

Different groups within a local community have not come to the table with equal power, and many of them often have known or cared little about the long-term consequences of environmental degradation. Without the threats of top-down regulations and lawsuits, powerful local groups often felt little need to engage with less powerful groups—they were perfectly happy to engage with industry instead. Romanticizing traditional, local resource management schemes and vilifying top-down government regulation ignores the history of environmental degradation that necessitated intervention in the first place.

The answer is not simply to give complete power to local communities but rather to institute a democratic process that creates a structure for useful conflict. A democratic process, the pragmatists argued, should both empower multiple voices and devise a method for negotiating the conflict those multiple voices soon lead to. For the pragmatist John Dewey, democracy was not merely a form of external government but "a body of tools and methods for undertaking the ongoing reconstruction of social life."[24] Because the world is constantly changing and public values are constantly shifting, the ways of providing "for the individual and common good have to be experimentally determined."[25] The neopragmatist philosopher Richard Rorty argues that institutions need to be viewed "as experiments in cooperation rather than as attempts to embody a universal and ahistorical order" and that pragmatism "tells us that we are going to have to work out the limits case by case, by hunch or by conversational compromise, rather than by reference to stable criteria." For Rorty, community, like democracy, is "something we make in the free interplay of diverse voices."[26]

What should the role of state power and federal land managers become in such a democratic process? How can adaptive management help create a process that enables what Rorty calls the "free interplay of diverse voices"? Federal managers can structure a process that enables different groups, with different amounts of power in a local community, to come to the table and be heard. They can force the powerful to listen to the powerless. They can get issues on the table that have been ignored for centuries. They can disrupt orthodoxy. All too often, of course, they do nothing of the sort, becoming part of the orthodoxy rather than its enemy. But democracy's promise lies in the process of getting people to the table and in giving new voices a chance to be heard.

The problem with this process is that it can heighten conflict, and few people like conflict. When only a few have a voice, and they all agree, management is much more efficient. Most managers prefer to avoid conflict and so suppress disagreements, but finding a process for hearing and negotiating conflict is critical in creating democracy. Dewey argued that the method of democracy

> is to bring these conflicts out into the open where their special claims can be seen and appraised, where they can be discussed and judged in the light of more inclusive interests than are represented by either of them separately. . . . But what generates violent strife is failure to bring the conflict into the light of intelligence where the conflicting interests can be adjudicated.[27]

Burying differences only worsens them; hiding conflict only prolongs it.

How can groups in a democracy best negotiate conflict? Why do some groups get stuck in orthodoxy and bitterness, yet others find a way to make progress? Examining the ways groups form their own identities offered Dewey and other pragmatists a way of approaching these differences. Dewey recognized that people within groups share values, beliefs, customs, symbols, assumptions, and expectations but are rarely able to think critically about these. Instead, groups usually naturalize their own values, making them seem inevitable, God-given, and beyond the reach of discussion or change. People often find themselves unable to change such naturalized perspectives without abandoning what they see as part of themselves. As the modern pragmatic philosopher James Campbell writes, "We hold on to sacralized elements of our past because to reject them is to reject ourselves."[28] For democratic processes to succeed, people need to be able to step out of their group's constructed view of itself and its past, and to recognize the validity of other groups' perspectives and values. They need to find a way to deconstruct their sacralized past with its sense of natural inevitability. Dewey believed that the critic's role was to help this process along. Critics should be people who are "rooted in the life of their community," who can see a community's "problems and possibilities of resolution and who try to bring this perspective before the public." Critics don't simply attack their society; rather, they appeal "from a narrow and restricted community to a larger one. . . . If successful, social critics make the community a richer place by causing it to confront and overcome the contradictions between its values and its current practices."[29]

How can this process work in adaptive management when different groups face conflicts over resources? In a fascinating study, the ethicist Paul B. Thompson examines the ways a pragmatic historical analysis can help resolve conflicts in water policy. Thompson argues that conflicts over water usually show certain predictable fissure lines. Property owners who hold existing riparian rights—in Malheur's case, ranchers such as Peter French—base their claims to water on a libertarian interpretation of property rights, believing that these rights give them authority to use their property however they like, as long as such use doesn't interfere with or harm others. In contrast, utilitarians in water conflicts—such as developers and progressive conservationists—believe that when ranchers' control of water produces less value than irrigation or reclamation would, the state is justified in taking over that water for irrigation. Finally, egalitarians in water conflicts, often environmentalists such as William Finley, argue for preserving some public good.[30]

In his experience with groups in conflict over limited water, Thompson found that providing each side with a theoretical justification for the policy they already favored served only "to freeze the actual political debate into stasis."[31] What worked far better than assigning ethical labels to fixed group identities was helping groups understand how fluid their own identities had been during their tenure in the landscape. As Thompson argues, a critic can show

> disputants how their opponents have used reasoning that they themselves might have used in a different situation to arrive at the opposing point of view. . . . It would be useful for each group to see themselves as part of the same community, at odds on a given issue, perhaps, but drawing from common moral traditions and headed towards a common future. Such a community might find political solutions that reciprocate each interest, even while they may demand compromise on the case at hand.[32]

A historical, pragmatic approach would find it useful to deconstruct the images of the rancher as a greedy cattle baron, of the irrigationists as out-of-touch technocrats, and of the environmentalists as malcontents. A pragmatic approach would show how they have had common interests and have been members of overlapping communities.

In the Malheur Lake Basin, disrupting settled notions of group identity can likewise be useful in resolving conflict, for it can show how members of overlapping groups have common interests in their shared place. People in the basin constructed their own identities in relationship to the riparian landscapes, partly out of the work they did, partly out of their conflicts with other groups, and partly out of the stories they told that defined their justification for being there. Yet these constructed identities shifted frequently, as the social and ecological landscapes shifted. The homesteaders who squatted on the contested lake bed had initially identified most strongly with the irrigation speculators, for both hated the cattle barons, believing ranchers had an unfair monopoly on land and water. Both groups believed that ranchers' refusal to build storage reservoirs that would control floods during the early spring and release water during the dry summer was an intentional waste of resources, a refusal to abide by the "duty of water."

To disrupt the political hold of ranchers over the basin's ecological riches, homesteaders initially joined with irrigators in an uneasy alliance. Yet soon after Peter French's murder, their progressive vision of social equity was hijacked by the new ranchers who replaced him—men such as William Hanley, who learned to shift their own identities, from ranching to irrigation speculation and development in an attempt to consolidate power in the basin. These hybrid ranchers and irrigators claimed the identity of conservationists even as they bitterly opposed the refuge.

In response, homesteaders soon broke rank with irrigation developers, joining with refuge management in 1922 to protect the refuge by blocking the state compromise that would have excluded both squatters and the refuge from control of water in the basin. And so, with the homesteaders' help, the federal government gained control of the basin in 1934. Ironically, in finally settling squatters' claims to the lake bed, the government paid the homesteaders enough for them to set themselves up as ranchers—the group they had so bitterly vilified. Conservationists, in turn, soon began to borrow elements of the identity of cattle barons to support their own dreams of an empire of ducks that was modeled after the mythic cattle empires.

What saved Malheur Lake from drainage was the willingness of the Otley family to align homesteaders with the refuge. On Steens Mountain eight decades later, the Otleys were once again willing to find common ground with their former enemies, so that negotiations over wilderness designations could move forward. Fred Otley—the former president of the Oregon Cattlemen's Association and descendent of the homesteader who had aligned with the refuge to block the irrigation/state compromise in 1922—helped negotiate the Resource Advisory Council's proposal to create the first cattle-free wilderness in the region. Few people believed that Otley had shifted his own identity from cattleman to radical environmentalist, but he realized ranchers had more to gain by compromising than by sticking to their guns. As this example shows, farmers, irrigators, ranchers, and federal employees all have shared and shifting identities. Long-term solutions that take these into account will have the most chance of success.

The secretary of the interior's threat to designate Steens Mountain a National Monument was a powerful motivator for groups to come to consensus, as were the threats of additional litigation from environmental groups. Even more important, the presence of a common enemy—Californians—helped unite ranchers and environmentalists in a shared identity as Oregonians who belonged to the basin. As much as ranchers and environmentalists distrusted each other, they distrusted Californians more. Developers were beginning to buy up inholdings on Steens Mountain, and both environmentalists and ranchers feared the area would be turned into another California and that its identity as a remote, wild, Oregonian place would be destroyed. The poet Gary Snyder (ironically, a Californian) argues that to make progress in protecting environments and rural communities, "we must learn to know, love, and join our place even more than we love our own ideas. People who can agree that they share a commitment to the landscape—even if they are otherwise locked in struggle with each other— have at least one deep thing to share."[33] In not wanting to let Steens Mountain turn into another California, ranchers and environmentalists came to know, love, and join their place even more than they loved their own ideas. Fear of a common threat is a powerful motivator for change.

Pragmatic adaptive management holds promise for the Malheur Lake Basin, but how useful is it for other places? The process certainly holds inherent dangers. Collaboration is painfully slow, and species can go extinct before decisions are made. Solutions come through compromises, and those compromises become possible when people are willing to shift their identities and find common ground with former enemies. But without strong enforcement of environmental laws, such compromises can prove dangerous for those members of the community lacking a voice: redband trout, wetlands, and riparian vegetation. Most people can agree that certain things should not be negotiable, but if pragmatic adaptive management gets bogged down in trying to negotiate these issues, the process will stall. For example, we cannot endlessly negotiate about what an endangered fish needs to survive; at some point we have to listen to the scientist who tells us fish need water, even if that biological fact narrows the possibilities open to resource users. Likewise, we have to listen to locals who tell us everyone needs to be included in decision making, even if managers find the process unwieldy.

Pragmatic adaptive management can work only if each group is willing to give up something for the benefit of the community. It cannot work when people become frozen in set positions, unwilling to budge toward any compromise. This can easily happen when imbalances in power make some groups feel they do not need to compromise—because they have nothing to lose by sticking to their fixed positions, or nothing to gain by negotiating. Threats of lawsuits and fears of the consequences of not negotiating are often necessary to keep the process viable. Adaptive management has the most promise on a local scale, not on a national or international scale, where finding common interests and shared identities as members of the same place becomes far more difficult.[34]

At Malheur National Wildlife Refuge, management's path is now as complex as the river's course once was. Legal battles constantly challenge refuge policy, much to the eternal frustration of staff who are trying to get their jobs done. But such outside influence is a good thing in the long run, however annoying it is from day to day. Without constant criticism and political pressures and court cases, refuge management would be far more efficient—and in the end, far more dangerous.

When managers work in isolation, they can come to operate with the ideological certainties that drove John Scharff's plans. Recent managers at Malheur have had a far more difficult time getting things done than did Scharff, for they have been bogged down in court cases, tied up in endless negotiations with different stakeholders, distracted by petitions to list native fish, and dragged into fights with hot-tempered neighbors. Although these are all enormous hassles, they enable federal agencies to chart a responsive course in a changing political, social, and ecological landscape. For example, the refuge is now paying attention to native fish, calling into question many decades of single-species management that benefited ducks but harmed much else. This change came about in large part because of the threat to list redband trout under the Endangered Species Act. Such legal threats have brought about new ideas, new conservation agreements, and new policies.

In his essay "Cowboy Ecology," environmental historian Donald Worster argues that the best grazing regimes over the long run have been the ones that had the most input from multiple members of the community. The same is surely true for refuge policy. Dealing with ranchers, environmentalists, county commissioners, Bureau of Land Management range conservationists, district court judges, tribal representatives, fisheries biologists, archeologists, and engineers makes a manager's life complicated. Trying to keep all these different stakeholders happy is an impossible task, but the attempt makes for an intelligent, evolving, policy—not a perfect policy, but one that can respond to change, just as a healthy river responds to changes. Ironically, what many managers dislike most about their job—the constant interruptions by stakeholders—may ultimately be what leads to the best refuge policy.

One hot summer day in 1996, I spent an afternoon walking through a creek just above Malheur National Wildlife Refuge. Forrest Cameron, then the refuge project leader, and Gary Ivey, then the refuge biologist, took me along a little cow trail through rabbitbrush and sagebrush that cut down in jagged Zs through the canyon into the creek. The trail led to a small gap where cows still had access to the water. The permittees had refused to maintain the fences that the Bureau of Land Management and refuge staff had placed to keep cattle from the riparian zones. Cameron and Ivey were in continual tension with the permittees, trying to avoid open conflict by going up there on days when they hoped the permittees would be in town.

We slipped down into the unfenced section of the riparian area where the creek bed was shallow and wide, and warm water ran across rocks coated with algae. Cow pats carpeted nearly the entire dry creek bed. Special permits allowed movement of cattle through the stream valley only in the spring, on their way up to summer grazing at higher elevations. But instead, permittees let their cattle linger in the riparian area. No cows were around when we were there, but we found shallow water, few willows, no currant bushes, no lush overhanging banks. A few alder hung on, their branches browsed off. Flies swarmed over the dirt. The side banks were sloughing off into the shallow creek.

Then we turned around, squeezing our way through the fence that marked the edge of the permittees' access to the creek, and found ourselves along a stretch of creek that had been protected from grazing for less than a decade. Already, willows and alders were so thick along the creek that we could not bash our way through. Grasses and sedges grew along the banks, in the new soil held in place by the willow roots. The sedges leaned over the channel, providing shade for the pools that were starting to form. The banks were no longer sloughing off into the creek as quickly, for they were now bound together by roots. The creek was beginning to heal: first the young willows found some protection from cattle teeth and hooves, and as they began to grow, their roots captured some of the sediment brought down by the spring runoff. Instead of clogging fish gills lower down in the stream's reaches, that sediment was now building up, providing more soil for more plants that caught yet more sediment—a positive feedback loop that could reverse the degradation of the creek.

A devil's advocate might ask, Which section of creek is really more natural? Why assume that a shallow, wide creek filled with algae and sediment and cow feces is any less healthy than a deep, narrow, winding channel shaded by willows and cloaked with sedges? Why does it matter that, in the fenced section, the water runs for a longer period each year, the banks hold during floods, roots and logs trap sediment, the creeks swarm with invertebrates, and a redband trout might be lurking in a pool? Why does it matter that the restored stream seems alive, rather than a cesspool for the passage of cow droppings? Why does it matter that the floodplain once again hosts a lush meadow, with willows and alder lining the channel? Why does it matter that the restoring creek may eventually gain access to the floodplain?

Richard White argues in *Organic Machine* that our long history of alterations along the Columbia River means not that the Columbia is dead but that it now speaks the story of a slow warm river of carp and shad, rather than a fast, cold, salmon river. But it is still natural, still alive, just a different nature. Can the same be said of the riparian areas, the little waters in the West? I think not. When you remove critical connections between the water and land, at first you may have a different stream. But eventually, you often have no stream at all. A stream may become intermittent, incising itself into the channel so deeply that the water table drops, and all that's left is a trace in wet years of the former stream. There is no natural essence of stream that can be separated out from the land and plants and soil that surround it and run beneath it. Without those connections, you have no stream. When those connections are restored, the stream can return.

Donald Worster, in his essay "A Country without Secrets," calls for environmental historians to study the processes of ecological adaptation, not just ecological destruction, in the American West:

> Most of that process of adaptation has been on an unconscious, unintended, often indirect level, creating a tangled web of nature, technology, and folk mentality; but there is also a story of conscious, *intentional* adaptation to tell. Now and then people did deliberately try to understand their ecological situation and developed explicit ideas about how to adjust their culture accordingly.[35]

Worster urges historians to pay attention to the lives of those people who came West and somehow adapted to limited water and limited resources.

In this book, I have looked for those who tried to adapt themselves to these watery landscapes. The story I have told has often been a tale of human folly and of decent people making decisions that led to unfortunate consequences for nature and culture both. But those are not the only stories to tell about human relationships with nature in the riparian West.

Different groups' responses to floods offers an example. Floods shaped, and continue to shape, human responses to riparian landscapes. Most people hate floods. Floods wipe out human signs of progress: they rip out roads, destroy crops, stink up basements, drown children. Floods are a slap in the face of human industry, and for many people in the Malheur Lake Basin, floods were a powerful source of frustration and anger. Yet some groups—the early ranchers who harvested native hay from riparian meadows and the refuge managers who have tried to mimic floods—have adapted to these disturbances. Rather than trying to regulate and engineer change out of existence, they learned to live with variability. Rather than trying to lock rivers and streams within their channels, they learned to take advantage of the richness offered by periodic chaos.

The other critical story I have traced in this book is that of connections. the complicated connections between water and land, and between human and nonhuman. What managers must contend with now at Malheur is a weird mixture of the past and present and the natural and artificial. As the ecologist James Young argues,

The vestiges of the past haunt the current managers of the riparian zone. We suffer from a lack of historical perspective when we contend with controversial practices—chaining juniper, restoring streams with instream structures, removing cattle from the creeks—without knowing the past. . . . A plant community is the product of how its predecessor was destroyed.[36]

We live in a world of ghosts that come back to haunt us when we try to fix other people's mistakes. We cannot find remedies without knowing what went wrong and why—not to lay the blame on the past, but to make sense of how we got to where we are.

During the summer months that I spent at the Malheur Refuge, my dog Juneau and I often walked just before dusk out into the marshes, along an access road surrounded by rushes and cattail and muddy banks. The bulrushes bent over where the muskrats pushed through them. Juneau crashed after the muskrats, while I slapped at the incessant mosquitoes. The vast bowl of the desert sky opened up above us. The waters, impounded behind dikes, lay still and silent and mirrored on either side of the canal. Cranes warbled. Coyotes howled, and my dog howled back. Even in a world profoundly altered by humans, wildness was all around us. Refuge staff had created those water systems not to increase agricultural production but to restore lands ravaged by drainage. Some critics claim that the artificial systems are little more than bathtubs or farm ponds. Yet, for all Malheur's artificial manipulations, it is still a wild place: wild in the profusion of vegetable and animal life buzzing and growing and burrowing and dying all around us, just outside our reach.

Malheur's history shows that although what people do profoundly affects nature, people can never control nature—a critical distinction. People try to create an artificial machine of water and land, but that machine soon swings out of their grasp. More and more in modern life, we fool ourselves into believing that human and ecological processes are separate. Urban life may make the connections between people and nature increasingly difficult to see, but that does not make them any less important. We are intimately connected to nature, and nature is intimately connected to us. Malheur's history reveals the tangled intimacies of human dreams and labors written upon the landscape, the messy connections between water and land and between human and natural.

NOTES

INTRODUCTION

1. Thomas et al., "Wildlife Habitats in Managed Rangelands"; Ohmart, "Historical and Present Impacts of Livestock Grazing," 246. Much of the ecological discussion that follows is indebted to Naiman, ed., *Watershed Management;* and Elmore and Kauffman, "Riparian and Watershed Systems," 212–31.

2. Hunter, Better Trout Habitat, 4.

3. Interview with staff biologist, Hart Mountain Antelope Refuge, June 1996.

4. Interview with Forrest Cameron, Project Director, Malheur National Wildlife Refuge, and Gary Ivey, Wildlife Biologist, Malheur National Wildlife Refuge. July 25, 1996. The "Annual Narratives" and the "Quarterly Narratives" from 1935–1996, stored in the Malheur National Wildlife Refuge Station Library, provide a good record of changing management practices.

5. USDI Fish and Wildlife Service, Malheur National Wildlife Refuge, *Annual Narrative* 1987, 2.

1 / RANCHERS IN THE MALHEUR LAKE BASIN

1. The literature on riparian ecology has mushroomed in the last decade. For a useful review of riparian function and ecology from a habitat perspective, see Hunter, *Better Trout Habitat;* and Gregory et al., "An Ecosystem Perspective of Riparian Zones." Malanson, *Riparian Landscapes,* offers a helpful overview of riparian function. Naiman, ed., *Watershed Management,* is an excellent introduction to riparian and river ecology. See also Thomas et al., "Wildlife Habitats in Managed Rangelands"; Ohmart, "Historical and Present Impacts of Livestock Grazing"; and Elmore and Kauffman, "Riparian and Watershed Systems."

2. French, *Cattle Country of Peter French,* 2, 42.

3. Shirk, *The Cattle Drives of David Shirk,* 128. For detailed descriptions of presettle-

ment conditions along the Blitzen River, see the useful reconstruction of General Land Office survey notes in Beckham, "Donner und Blitzen River, Oregon."

4. Naiman et al., "Ecosystem Alteration of Boreal Forest Streams by Beaver *(Castor canadensis)"*; Naiman et al., "Alteration of North American Streams by Beaver."

5. Ogden, *Snake Country Journals 1827–28 and 1828–29,* xiv, n. 1.

6. Ibid., 98.

7. Ogden noted the refusal of the Native Americans to trap the streams dry, writing on September 13, 1827, "If the Cayouse will not ruin the beaver in their own lands, we must for them at least assist to diminish the number, and if we do not others probably will for us, and at no distant period" (ibid., 5) and, several days later, "Indeed the once famed Snake Country for beaver is a ruined one now, and granting it were allowed repose, which few on either side concerned are willing to give, it would at least require four years to recruit. I for one am willing to give it; for my present state of health I feel I require repose myself" (ibid., 17).

8. Elmore and Beschta, "Riparian Areas," 260.

9. Ogden, *Peter Skene Ogden's Snake Country Journals, 1824–5 and 1825–6,* entry for February 3, 1826, p. 123.

10. French, *Cattle Country of Peter French,* 43.

11. Robbins, *Landscapes of Promise,* 157; Lo Picollo, "Some Aspects of the Range Cattle Industry," 7–22.

12. Blitzen Valley Land Company, "General Description of the Property," 1.

13. Robbins, *Landscapes of Promise,* 159.

14. Simpson, *Community of Cattlemen,* 5–6. For example, John Devine controlled 840 square miles of range, shallow marshes, and desert lake bed east of Steens Mountain.

15. Ibid., 2.

16. Oppenheimer, "Rancher Subsidy."

17. Testimony from court cases in the 1920s attests to Peter French's fencing program. One local man, Maurice Fitzgerald, stated in court, "Of course. He had a lot of stone fences around the Diamond ranch. He put up gaps in the rimrock; he constructed large heavy stone walls and put gates leading into those parts of the ranch where those stone walls were; he had gates through them" (Fitzgerald, "In the Matter of the Determination," 389.

18. "In March of 1887, the *Silver State* noted that John Devine was on his way to the United States Court in Portland to answer charges of illegally fencing the public domain and that in September he moved a fence to comply with the orders of the Interior Department. By 1889–90 the General Land Office noted the removal of the offending Miller & Lux fence while two years earlier, in 1887, the *East Oregon Herald* announced that much land, previously under fence, was now open for settlement" (Lo Piccolo, *Some Aspects of the Range Cattle Industry,* 38). On fencing conflicts in the basin, see also Simpson, *Community of Cattlemen,* 59.

An article in the local newspaper, the *Harney Valley Items,* reveals some of the tensions over fencing: "ALL THE FENCES MUST GO: In Western Harney and Crook Counties, 30,000 to 40,000 acres will be opened. Dispatch from Prineville in the Oregonian said: The 'order for the removal of fences from Government land is causing considerable indignation in the Eastern part of the county.' On overgrazing in other places, 'There are whole ranges in Texas, New Mexico, and Arizona, once rich beyond belief, that are completely deserted and given over to the desert'"(Feb. 7, 1903).

19. Talbot, "P Ranch," 24. By 1887 French was castrating males and enforcing tight control of bull quality, and running more than forty-five thousand head of cattle—methods derived from Spanish heritage that were diffused along the great cattle trails north to Canada (Simpson, *Community of Cattlemen*, 4; for a much fuller discussion, see Jordan, *North American Cattle-Ranching Frontiers*).

20. Eastern methods of cattle raising eventually became dominant in the region (Simpson, *Community of Cattlemen*, 4).

21. Ortega, "Testimony," 322, 360.

22. Blitzen Valley Land Company, "General Description of the Property," 3.

23. Ortega, "Testimony," 318.

24. Fitzgerald, "In the Matter of the Determination," 392.

25. Myron Angel, cited in Lo Picollo, *Some Aspects of the Range Cattle Industry*, 77.

26. In reply to a lawyer's question about Peter French's putting up hay in 1877, Prim Ortega replied, "Yes, he put up some but he did not put up very much, you know, at that time, they did not have to, you know, stock never got poor. There was lots of grass all over the range; but he put up a little, you know. Enough if he wanted to keep up stock, to work stock, or something like that, he would feed; he did not have to cut much hay at that time" (Ortega, "Testimony," 336).

27. Lo Picollo, "Some Aspects of the Range Cattle Industry," 24.

28. Ibid., 25. Charles Lux was the Lux in Miller & Lux. See also Brimlow, *Harney County Oregon*, 8.

29. The Silvies River flowed into a low basin, with nearly "700,000 acres of flat alluvial soil" forming excellent potential agricultural lands. The USGS geologists F. F. Henshaw and H. J. Dean noted in 1915 that the lower lands were "more or less swampy and for years have been utilized as hay flats. During the spring months, when water is abundant, small-rock-and-brush dams sufficiently high to cause the streams to overflow their banks are constructed at convenient points. Much of the meadow lands may thus be covered by a foot or more of running water during five or six weeks of the season. As the stream flow diminishes, the swamp and hay lands gradually drain until the middle of July, when they are sufficiently dry to permit the cutting of hay" (Henshaw and Dean, *Surface Water Supply of Oregon*, 74).

30. Miller, letter to ranch manager H. N. Fulgham, Dec. 28 1891.

31. For a wonderful description of the effort involved in flood irrigation, see deBuys and Harris, *River of Traps*.

32. Treadwell and Rand, "Brief in Support of Claims," 78–79.

33. Miller, letter to Fulgham, Dec. 28, 1891.

34. James Brandon, testimony, in Treadwell and Rand, "Brief in Support of Claims," 95.

35. Miller, letter to Fulgham, Jan 30, 1892.

36. Treadwell and Rand, "Brief in Support of Claims," 110.

37. Vander Shaff, "Final Report," 14, 36.

38. Fiege, *Irrigated Eden*, 42–80.

39. Lewis, "Irrigation in Oregon," 29.

40. Simpson, *Community of Cattlemen*, 18.

41. Robbins, *Landscapes of Promise*, 159.

42. Simpson, *Community of Cattlemen*, 40.

43. Ibid., 24, 27; Lo Picollo, "Some Aspects of the Range Cattle Industry," 84.

44. George Hibbard remembered in an interview that the Indians near Lakeview used mullein stocks, the leaves of which contain a chemical that acts like rotenone, to numb or suffocate the fish, which could then be easily caught (Oral History tape #191, 1975, Harney County Oral History Project, Burns Library files, Burns, Oregon), 3.

45. Indian agent Dennison, Commission of Indian Affairs, 1857, cited in Steward and Wheeler-Voegelin, *The Northern Paiute Indians*, 47.

46. Beatrice Blyth Whiting, "Paiute Sorcery," (1950), 19–20. Couture, "Recent and Contemporary Foraging Practices," 31.

47. James Dale Wilde, "Prehistoric Settlements in The Northern Great Basin: Excavations and Collections Analysis in the Steens Mountain Area, Southeastern Oregon" (1985), 80–81.

48. Ibid.

49. Wayne Elmore, interview with author, June 15, 1997.

50. The Paiute Culture Committee today maintains that riparian burning is still important to the tribe; for example, one elder argued in 1997 that natural fires should burn through areas of critical environmental concern for better root production (interview with Carla Burnside, 6/23/97).

51. Frémont, *Report of the Exploring Expedition to the Rocky Mountains*, 593.

52. Ibid., 587, 599.

53. Captain Henry D. Wallen, "Report of Expedition," in *Report of the Secretary of War* (1859–1860), 16.

54. Currey, *Report of the Adjutant General*, 11.

55. Currey, entry for Oct. 20, 1865, in *Report of the Adjutant General*, 81–2.

56. General George Cook, *Autobiography*, 142, 145.

57. Not until Sept. 12, 1872, by executive order of President Grant, were the Paiute granted the Malheur Reservation of 1,778,560 acres. But the Senate never ratified the treaty,

which meant that the government took the lands into possession "without formal relinquishment by the Indians."

A. B. Meacham wrote in 1871 in "Report of the Commissioner of Indian Affairs" (p. 304), "If our Government intends to be just and uniform in its treatment of Indians, these people should be provided for without delay, although they may not be the possessors of enough political power to secure to them the consideration of local politicians, they, at least, as original inheritors of the soil, have a 'God-given "right" to life, liberty, and the pursuit of happiness'; . . . As an officer of the United States Indian Department, I demand, in behalf of the Indians of Oregon, that their rights be regarded, and justice done them in some way, either by paying them for their lands, or allowing them to locate homes in common with the white man who is making the Indian country so valuable" (reprinted in Meacham, "Notes on Snakes, Paiutes, Nez Perces at Malheur Reservation").

As the Bureau of Indian Affairs wrote in 1963, "The Federal Government . . . had extended its authority without formal purchase over the territory of the 'Western Shoshoni' and included within it the northern part of the lands occupied by the Northern Paiute tribes, assuming 'the right of satisfying their claims by assigning them such reservations as might seem essential for their occupancy and supplying them in such degree as might seem proper with necessaries of life.' The Paiute did not formally relinquish their lands and they refused to stay confined on the reservation." USDI News Release, Aug. 6, 1963, Bradley-Interior 4306, Burns tribal archives.

58. Steward and Wheeler-Voegelin, *The Northern Paiute Indians*, 189.

59. William Rinehart, "Report of the Commission of Indian Affairs, 1876" reprinted in Meacham, 271–73.

60. Ibid., 274–80.

61. Interview with Linda Reed, archaeologist for the Burns Paiute tribe, July 1, 1997.

62. Rinehart, "Report of the Commission of Indian Affairs, 1878," 285.

63. Ibid.

64. About one hundred Winnemucca Paiute did not participate in the uprising. The Paiute knew they had no chance of winning a war against the whites, but they felt that continued subservience meant a slow death. When asked why he had joined the uprising, Chief Egan said Rinehart "had lied, cheated and robbed them and at times refused to issue subsistence when he had ample supplies on hand and 'they might as well die fighting as to be starved to death,' that while they did not expect to whip the whites, he would fight as long as he could" (ibid., 284).

65. Ibid., 287.

66. Ibid., 291. Rinehart complained in his 1879 report, "About the date of the removal of the Indians to Yakama in January last, it was announced in the public press, under the caption, 'General Howard interviewed,' that 'the Malheur Reservation must and shall be broken up.' The same announcement was reported to have been made by military officers

at Camp Harney. . . . And encouraged by these announcements, stock-men and settlers immediately went upon the reserve with their herds and occupied the most valuable portion of the agricultural and meadow lands. These trespassers have not yet all been removed."

67. Beckham, "Donner und Blitzen River, Oregon," 11–12.

68. Robbins, *Landscapes of Promise*, 160, citing Simpson, *Community of Cattlemen*, 18, 29–30.

69. Lo Picollo, "Some Aspects of the Range Cattle Industry," 84.

70. Simpson, *Community of Cattlemen*, 41.

71. Ibid., 31. Although many of the small owners sold out to the big operations during the 1880s, many of them took that capital and got title to other lands within the county.

72. Lo Picollo, "Some Aspects of the Range Cattle Industry," 66.

73. Simpson, *Community of Cattlemen*, 45.

74. Shirk, *The Cattle Drives of David Shirk*, 31.

75. Lo Picollo, "Some Aspects of the Range Cattle Industry," 65.

76. Blitzen Valley Land Company, "General Description of the Property," 2.

77. Ortega, "Testimony," 362.

78. "Although fraudulent entries usually passed unnoticed by the Federal Government this was not always the case. In 1883 the General Land Office canceled the desert land entries of James A. Jennings, J. M. Deadman, Isaac W. Laswell, John J. Hallett, and Peter French, all of Lakeview Land District in Oregon. The decision was based on two facts. First, the land was non-desert in character and second, Peter French employed all the men and they made the entries in his behalf. French appealed the decision but the Land Office remained firm. This case, however, was an exception rather than a rule" (Lo Picollo, "Some Aspects of the Range Cattle Industry," 66–67).

79. Ibid., 54. For a useful description of federal efforts to distribute wetlands, see Vileisis, *Discovering the Unknown Landscape*, 72–91.

80. Lo Picollo, "Some Aspects of the Range Cattle Industry," 54.

81. Jackman and Scharff, *Steens Mountain*, 37.

82. Ibid., 37.

83. Lo Picollo, "Some Aspects of the Range Cattle Industry," 54. The Oregon Supreme Court rendered a peculiar decision in 1874, saying that the state had the right to dispose of these lands before the secretary of the interior had formally granted title to them to the state.

84. Jackman and Scharff, *Steens Mountain*, 37.

85. Waring, "Geology and Water Resources," 16.

86. William Robbins wrote, "All of those outfits were corporate operations, with the owners representing outside capital, especially from California. The famous Miller and Lux outfit from that state eventually gained control of the Thomas Overfelt holdings on the Malheur River in 1875 and then added the Todhunter and Devine operations when

the latter went bankrupt in 1889. With the Devine acquisition, Miller and Lux reorgan-ized into the huge Pacific Livestock Company" (Robbins, *Landscapes of Promise*, 161).

A few cattle operations gained tremendously from these acts. One Harney County cattle baron, W. B. Todhunter, gained title to 40,332.13 acres, including large tracts on the north shore of Malheur and Harney Lakes, sections east of Malheur Lake, and sections east of Steens Mountain. Land speculators also took advantage of the Swamp Land Act, and made huge profits by selling off to ranchers. For example Henry C. Owens of Eugene, Oregon, applied for all swamp lands between Beatty's Butte and Steens Mountain; the state granted him title to 484,059.85 acres between July 26, 1880, and January 20, 1882. Owens sold 21,942 acres to Todhunter, and much more land northeast of Malheur Lake to another cattle company, Riley and Hardin. The state eventually rejected most of this claim. In 1889 the Pacific Livestock Company bought the holdings of Todhunter and Devine, which included large tracts of swamp land (Lo Picollo, "Some Aspects of the Range Cattle Industry," 61).

87. Talbot, "P Ranch," 22.

88. Lo Picollo, "Some Aspects of the Range Cattle Industry," 63.

89. *Harney Valley Items*, April 11, 1889. Throughout Oregon, people began to see the cattle barons not as civilizers but as despoilers, alien monopolists who stole good land from decent families. In his inaugural address of 1887, Governor Sylvester Pennoyer lamented the effects of the Swamp Land Act: "The gift of the General Government of March 12, 1860 . . . was a Greek gift. The result of that gift has been that some of the fairest and most productive portions of our State, susceptible of supporting a large population, have been monopolized by a few individuals; immigrants that would have helped to build our free institutions have been turned away; and a few cattle barons claimed the soil. . . . A thrifty yeomanry is a far richer endowment to the State than a few thousand dollars in our treasury, as the price of turning large areas of our most valuable lands over into the possession of a few alien stock raisers" (Lo Picollo, "Some Aspects of the Range Cattle Industry," 58).

90. *Harney Valley Items*, Feb. 6, 1890.

91. *Harney Valley Items*, Feb. 16, 1887.

92. Simpson, *Community of Cattlemen*, 59. For example, along the Silvies River in the north end of the basin, debates concerning the nature of the land continued after Henry Miller and other partners in the Pacific Livestock Company bought the proper-ties of Todhunter and Devine. By 1891 the secretary of the interior rendered a decision in favor of the cattle baron, and the courts rejected the settlers' claims. Nevertheless, in his report of 1885–86, "The Secretary of the Interior, L. Q. C. Lamar, reported that some states allowed the sale of presumptive swamp lands at low prices before authorized selec-tions had been made and actually made selections upon the statements of the applying purchaser" (USDI *Report of the Secretary of the Interior 1885–88*, quoted in Lo Picollo, "Some Aspects of the Range Cattle Industry," 62).

93. Robbins, *Landscapes of Promise*, 161–62.

94. For an excellent study of overgrazing in a similar ecosystem, see Young and Sparks, *Cattle in the Cold Desert*.

95. Robbins, *Landscapes of Promise*, 161; quoting Simpson, *Community of Cattlemen*, 39–40. Robbins adds, "The center for livestock shipments subsequently shifted eastward to Huntington and Ontario on the Snake River when the Union Pacific completed its connecting route into eastern Oregon in 1884. The Oregon Short Line represented a shift in the cattle market to Chicago rather than Winnemucca and its more regional market in San Francisco. In effect, the movement of cattle eastward to Chicago linked south-eastern Oregon's grasslands to consumers in America's expanding industrial heartland."

96. "The ranch left a private, and probably accurate, estimate for 1884 and 1885 of 43,786 and 45,066 head respectively. A rough estimate of 30,000 to 45,000 head through-out the history of the ranch might be accurate, if not slightly conservative" (Lo Picollo, "Some Aspects of the Range Cattle Industry," 29.

97. Gordon, "Report on Cattle, Sheep, and Swine," 107, 124, 128.

98. Ibid., 107, 126. Gordon added, "The best stock ranges are in the eastern part and in the southeast corner of the county, especially those in the Malheur Indian reservation and about Stein's [sic] mountains. . . . Fencing has begun on the best conducted ranches in the Stein's mountain region. On the Glenn and French ranches there are over 30,000 acres inclosed [sic], the 'rim-rocks' greatly aiding this. There are instanced in this county where, by a comparatively small use of fence, 4 or 5 miles of continuous inclosure have been constructed. Where wood is used entirely, and it grows in the neighborhood, fenc-ing costs about $225 per mile."

99. Ibid., 128.

100. Lo Picollo, "Some Aspects of the Range Cattle Industry," 77.

101. Ibid., 106.

102. Simpson, *Community of Cattlemen*, 40; quote from Lo Picollo, "Some Aspects of the Range Cattle Industry." In 1880, what became Harney County (then part of Grant County) had 13,743 sheep and 46,250 head of cattle on the tax books; with 40 stockmen recorded in the census. By 1890 sheep had increased to 56,698, cattle to 27,328, and live-stock producers to 344. The census, however, does not reveal most of the cattle, accord-ing to Simpson. In the assessment rolls of Harney County of 1885, cattle were counted at 52,835 head (most of which never appeared in the census), and sheep at 61,470; the five largest livestock owners controlled 49 percent of all livestock, and the three largest cat-tle operations owned 60 percent of the cattle.

103. *East Oregon Herald*, Aug. 22, 1889, cited in Lo Picollo, "Some Aspects of the Range Cattle Industry," 108. See also *East Oregon Herald*, Aug. 1, 1889: "Meadow valley in Harney Valley perfectly dry and will yield little or nothing, while most of the grain fields look as if they had been harvested last year."

104. *East Oregon Herald*, Sept. 2, 1889. "Owing to the White Horse ranch being dried

out 3,000 cattle are being placed on the Harney ranges. . . . Waters on the Red-S lands are reported all dried up, and it is feared Silvies will be dry, too, if the present unconquerable dryness continues."

105. *Harney Valley Items,* Jan. 23, 1890; Jan. 30, 1890.

106. Simpson, *Community of Cattlemen,* 134–45.

107. Not all sheep were part of transient operations; about twenty sheepherders were resident in the county and were among the oldest settlers, people of property with a stake in the area. The tensions in Harney County were between taxpayers and transients, not between cattle ranchers and sheep operators per se. If one was a sheepman, that was fine as long as one owned land, paid taxes, and took part in the community (ibid., 134).

108. F. C. Lusk, letter to editor, *Harney Valley Items,* 1902.

109. Waring, "Geology and Water Resources," 16.

110. Griffiths, *Forage Conditions,* 28. Griffiths estimates between 450 and 1,000 animals per square mile on the mountain; 73 flocks on the mountain, each with at least 2,500 sheep, for four to five months; or 820,000 sheep months on the mountain that year.

111. Ibid., 29.

112. Ibid.

113. Ibid., 27, 56.

114. Ibid., 22, 23.

115. Ibid., 26.

116. Ibid., 56.

117. Ibid., 57.

118. Simpson, *Community of Cattlemen,* 58.

2 / CONFLICTS BETWEEN RANCHERS AND HOMESTEADERS

1. Simpson, *Community of Cattlemen,* 55.

2. "The cattle men are nearly all nonresidents; the cattle are sold out of the county and the money used elsewhere; all the benefit the county derives is the presence of the few men needed to care for the stock and the taxes upon the cattle, greatly undervalued both in quantity and price. Todhunter & Devine have 40,000 head of cattle; French & Glenn, 30,000; Thomas Overfelt, 30,000; Riley & Herdin, 25,000; C. G. Alexander 15,000; Sweetzer Bros., 25,000; George W. Mapes, 10,000; Wm. Hydespath, 6,000. Without counting the many smaller bands, this gives a total of 181,000, and yet the total number assessed for the year 1883–84 was but 74,611" (*West Shore,* Feb. 1885, 43).

3. *Harney Valley Items,* Aug. 3, 1901.

4. Simpson, *Community of Cattlemen,* 60.

5. Ibid. Mark Fiege reminded me that Matt Klingle's Ph.D. dissertation, "Urban by Nature: An Environmental History of Seattle, 1880–1970," deals with a similar conflict over tidewater lands in Seattle.

6. "'A great change is taking place in the physical condition of our section of coun-try,' said State Senator Cogswell of Lake County to a reporter of the Oregonian last week. 'The water of many of the lakes is subsiding, due in a measure to drawing it off for irri-gating purposes, and also, to natural causes. Not over four square miles of the original bed of Warner Lake is now covered with water. In 1865 there was seven feet of water on it. . . . In Warner valley, a few days ago, 300 tons of hay were burned; in 1874 the spot where the fire took place was surveyed as Warner Lake. Goose Lake has subsided five feet since 1869. . . . Lake Malheur of Harney county registers eight feet lower than at any period within the memory of the oldest inhabitant'" (*East Oregon Herald*, March 28, 1889).

7. When the editor asked what would happen to farms when the lake overflowed again, T. V. B. Embree replied, "Now in case of another great and continued overflow let a number of settlers take desert claims along this plain, and dig a canal from the lake to the river and thus kill two birds at one throw—reclaim the desert, also, the swamp lands. . . . Parties have questioned me continually about this country overflowing again. People may just understand at first, and once for all coming time that they must take land that will be subject to annual overflows, or else they must take land that is never overflowed. Which do you want? If you want the first, the lake country is as good, if not better than any. . . . It is simply soil with a luxuriant vegetation growth upon it. With industry and enterprise it will be the best land in Harney valley" (Embree, letter to the editor, *East Oregon Herald*, Oct. 3, 1889).

8. W. A. J. Sparks, the reform-minded U. S. land commissioner, moved to disallow all swamp claims in Oregon; indictments were brought, but few claims reverted to the government (Simpson, *Community of Cattlemen*, 77).

9. Steinberg, *Slide Mountain*, 7ff.

10. Pisani, *To Reclaim a Divided West*; idem, *Water, Land, and Law*. See Cronon, "Landscapes of Scarcity and Abundance," 615.

11. Cronon, "Landscapes of Scarcity and Abundance," 616.

12. Ibid.

13. Hundley, *The Great Thirst*, 73.

14. Donald Pisani's work on miners shows a somewhat more complicated situation: although large mining companies pushed for prior appropriation, many small miners defended their riparian rights (Pisani, *To Reclaim a Divided West*).

15. *East Oregon Herald*, Feb. 5, 1889.

16. *San Francisco Examiner*, Feb. 6, 1889.

17. *Harney Valley Items*, April 8, 1889. Fights developed in editorial columns over the justice of arson. The *Grant County News* (a paper of townspeople, one that rarely sup-ported homesteaders) editorialized on Feb. 6, 1890, "The anarchists have been accom-plishing their hellish designs in Harney county again. What punishment is too severe for a fiend who will deliberately deprive poor dumb beasts of food for the sake of wreaking

vengeance on a man who has incurred his enmity? . . . The sooner Harney county breaks up their vile clan the better it will be for her good name and prosperity."

18. French, *Cattle Country of Peter French*, 153–56; Simpson, *Community of Cattlemen*, 80.

19. *Harney Valley Items*, Sept. 28, 1901.

20. *Burns Times-Herald*, August 23, 1902; July 13, 1901.

21. *Harney Valley Items*, June 23, 1903; Oct. 5, 1901.

22. Michael Robinson, *Water for the West*, 10.

23. *Harney Valley Items*, June 1, 1901.

24. Michael Robinson, *Water for the West*, 9.

25. Ibid.

26. Ibid., 13.

27. William Smythe, cited in ibid.

28. Pisani, "Beyond the Hundredth Meridian."

29. In 1887 this collective vision found expression in California's Wright Act, which created the collectivist institution of irrigation districts. Districts could take collective control of their water rights if two-thirds of voters approved a bond measure that used tax revenues to support irrigation investments (Cronon, "Landscapes of Scarcity and Abundance," 617). The hope among many proponents was that this would, in Norris Hundley's words, "foster community values, promote small family farms, and curb the monopolistic excesses produced by the rampant individualism of California's pioneer capitalists" (Hundley, *The Great Thirst*, 98). In *To Reclaim a Divided West* and *Water, Land, and Law in the West,* Donald Pisani explores the growing tensions between America's distrust of federal monopolization of natural resources and the West's need of federal help in irrigation.

30. As one of the early Oregon Desert Land Board reports stated, the withdrawal of lands would generally be made "by the state upon the request of some promoter who discovers a project, and who later contracts with the state for its reclamation without expense to the state. The promoter receives his pay by selling the lands to settlers after reclamation, at the price of the lien on each tract, which lien has been created by the state in favor of the contractor as an inducement to undertake construction" (State of Oregon, "Report of the Desert Land Board Relative to the Reclamation of Desert Lands Granted to the State Under the Provisions of the Carey Act," 1st Biennial Report [1911], 3).

31. State of Oregon, "Report of the Desert Land Board," 2nd biennial report, 3.

32. Lewis, "Irrigation in Oregon," 33.

33. State of Oregon, "Report of the Desert Land Board," 6th biennial report, 7.

34. Michael Robinson, *Water for the West*, 13. For example, the first biennial Desert Land Board Report, of 1911, noted, "The State was endeavoring to cancel in the courts, its contract with the Columbia Southern Irrigating Co. which had sold water rights far

in excess of the available water supply. This suit was decided adverse to the State. The old law placed the responsibility of supervising reclamation work upon the State Land Board. Having no funds, or experience in this class of work, the Board naturally assumed that the federal government, through its corps of inspectors, would look after the details. Plans and contracts were approved by the State mainly to get them before the department for inspection. To remedy these early mistakes has been difficult and in some cases impossible" (State of Oregon, "Report of the Desert Land Board," 1st biennial report, 4–5). Two years later, the report noted, "The parties initiating the projects did not seem to realize the importance of obtaining complete and reliable information relative to water supply, duty of water and cost of reclamation. The result was that the amount of available water was grossly over-estimated, or the amount necessary to reclaim the land under-estimated, or the cost of applying the water greatly under-estimated" (State of Oregon, "Report of the Desert Land Board," 2nd biennial report, 3).

35. As Robinson argued, "Strong opposition arose in Congress from eastern and midwestern representatives who advanced anti-reclamation arguments later echoed throughout the twentieth century: farmers were already producing surpluses; it would be unfair to aid western irrigators at the expense of eastern farmers; future generations should not be burdened by debts for water projects; opening lands was a government subsidy to the few; and Federal reclamation violated State rights and was therefore unconstitutional.

"[Francis Griffith] Newlands and supporters of his bill developed social and financial arguments in favor of enactment: it would extend the frontier and relieve population pressures and overcrowded cities; reservoirs would enhance conservation; irrigation crops would be consumed locally; the program would be self-liquidating; new markets would be created for eastern manufactured goods" (Robinson, *Water for the West*, 14).

36. Simpson, *Community of Cattlemen*, 95.

37. Cronon, "Landscapes of Scarcity and Abundance," 619.

38. Worster, *Rivers of Empire*, 7.

39. Donald Worster, "The Challenge of the Arid West," National Humanities Center website, <http://www.nhc.rtp.nc.us:8080/tserve/nattrans/ntwilderness/essays/aridwestd .htm> (no date, but after 2000).

40. Hundley writes, "Central to this story, then, is the appearance of a new kind of social imperialist whose goal was to acquire the water of others and grow at their expense. . . . At the same time, this is a story of extraordinary feats of fulfilling basic social needs when communities mobilized and focused their political energies on providing abundant clear water to multitudes of people who clearly wished that to be done" (*The Great Thirst*, xv).

41. Pisani, *To Reclaim a Divided West*.

42. *Burns Times-Herald*, August 23, 1902; *Harney Valley Items*, June 15, 1901.

43. Bastasch, *Waters of Oregon*, 43. For a brief review of water rights laws, see Cronon, "Landscapes of Scarcity and Abundance," 616. In California, the *Lux* v. *Haggin*

court decision of 1886 had held that riparian rights did come with the sale of land, unless someone else already had appropriated water rights before the sale; the California Doctrine of water use tried to balance riparian rights and appropriative rights. Large ranchers favored riparian rights, but others—such as miners, factory owners, and irrigators—objected, believing that their water use was more active, more interventionist, more beneficial to progress. As William Cronon argues, the California Doctrine recognized that "different users of water had different needs," which meant that different ways of conceiving of legal property rights to water might be necessary. In contrast, the Wyoming Doctrine took the concept of beneficial use to an extreme, when the Wyoming constitution claimed title to all water and "retained the power to alter any existing private rights or appropriations that did not serve the public interest" (Cronon, "Landscapes of Scarcity and Abundance," 616).

44. Mandaville, "A Swamp in the Desert," 73; Bastasch, *Waters of Oregon,* 44.

45. Rights were assigned only for the physical control of water, and only when that control of water is designed with an end toward the use of water. So, for example, one needed no water right for livestock to drink from a stream, since no physical control is involved. Likewise, one needed no permit to pump groundwater out of a quarry, since the intent was not to make beneficial use of the water, but rather to get rid of it (Bastasch, *Waters of Oregon,* 44, 51).

46. Ibid., 48.

47. State Engineer 1910, *Third biennial report of the State Engineer to the Governor of Oregon for the period beginning December 1, 1908, ending November 30, 1910,* cited in Bastasch, *Waters of Oregon,* 44.

48. Lewis, "Irrigation in Oregon," 27. For a fascinating examination of related questions in Idaho, see Mark Fiege, *Irrigated Eden,* 171–202.

49. Lewis, "Irrigation in Oregon," 28.

50. Ibid.

51. Ibid., 240–41.

52. Ibid.

53. Martin, "Nevada," 89, 90.

54. Ibid., 92.

55. Ibid., 91.

56. Cronon, "Landscapes of Scarcity and Abundance," 616–17.

57. Hawley and McConnell, "Contest No. 230," 248.

58. Fiege, *Irrigated Eden,* 9.

59. Simpson, *Community of Cattlemen,* xi.

60. Ibid., 164.

61. Jackman and Scharff, *Steens Mountain,* 164.

62. L. Alva Lewis, "Report on Conditions," 3, 14.

63. Lo Picollo, "Some Aspects of the Range Cattle Industry," 62.

64. Simpson, *Community of Cattlemen*, 107.

65. Jackman and Scharff, *Steens Mountain*, 164.

66. Quoted in Friedman, *The Other Side of Oregon*, 174.

67. As the local historian Ralph Friedman wrote, "By 1930 he had seen the rise of corporate agriculture and the decline of the small farmer. He was disturbed by the first and bothered by the latter, but his social consciousness had been developed long before. A liberal in a conservative business, he never forgot his own lean times, and he never let his wealth get in the way of his vision. He was sympathetic to labor, sided with the Indians, espoused women's rights, was nauseated by bigotry of any kind, championed conservation, looked upon World War I as 'commercial' in essence, had the populist contempt for the 'money lenders,' and 'preferred conciliation to violence'" (ibid., 174).

68. Simpson, *Community of Cattlemen*, 95.

69. Ibid., 94.

70. *Harney Valley Items*, June 27, 1903.

71. Ibid., Aug 29, 1903.

72. "A few years ago the United States Reclamation Service made preliminary surveys of two irrigation projects within the Harney basin. Of these, the Harney project embraced lands in the northern part of Harney Valley, while the other, the Silver Creek project, contemplated the irrigation of the bench lands along the stream above Riley.

"Concerning the larger project, that in the Harney Valley, it is stated in the Third Annual Report of the Reclamation Service that in 1904 there remained about 70,000 acres of unpatented land in the valley, and this had been selected by the State, under the Carey Act, for the Harney Valley Improvement Company. Measurements from May, 1903, to May, 1904 inclusive, showed that 250,000 acre-feet of water was discharged by the Silvies River into the valley during that period. Much of the discharge is spring flood water, and it has been estimated that fully one-half the total amount can be stored at comparatively small expense at an excellent reservoir site at the head of Silvies Valley, about 20 miles north of Harney Valley. After making due allowance for the fact that the season of 1903–4 was one of unusual precipitation it was assumed that there is sufficient water to irrigate 40,000 acres, and that the necessary storage and distribution work can be constructed at a cost not to exceed $20 an acre for the land irrigated. But while the present method of irrigation by flooding is practiced, in which from two to five times the amount of water is used than would be necessary if it were properly distributed, it is doubtful if in seasons of average precipitation there is more water than is required to satisfy present claims. In June, 1904, board of consulting engineers examined the project and recommended that it be abandoned, since the water supply would be found inadequate" (Waring, "Geology and Water Resources," 54).

73. Harney Basin Development Company, "Brief," 141, 139.

74. Simpson, *Community of Cattlemen*, 96.

75. Ibid., 95, quoting the *Harney Valley Items*, July 25, 1903.

76. Harney Basin Development Company, "Brief," 134–45.

77. Ibid., 139.

78. Ibid.

79. John T. Whistler, letter to William Hanley April 2, 1910. Hanley Company Papers, Oregon Historical Society MSS 378 Box 4.

80. John T. Whistler, letter to William Hanley, July 1, 1910, Hanley Company Papers, Oregon Historical Society MSS 378 Box 4.

81. Harney Basin Development Company, "Brief," 2693–94.

82. Ivey, "History of Riparian Habitat," 2. Ivey notes that channelization "changed the water regime . . . eliminating the curving oxbows of the river channel from the river's influence . . . [leading to] extensive reduction of riparian habitat along the old water-courses and some new development of willow communities along irrigation facilities."

83. Finley, "Malheur, The Unfortunate," 4.

3 / BUYING THE BLITZEN

1. Mitman, *Reel Nature*, 96–97. On Finley's life and work, see Mathewson, *William Finley.*

2. Finley, "On the Trail of the Plume Hunters." See also Jennifer Price's discussion of conservation and the feather trade in "When Women were Women," 57–110.

3. L. Alva Lewis, "Report on Conditions," 7.

4. Ibid., 12.

5. Sharp, *Where Rolls the Oregon,* 96.

6. Ibid., 94 (emphasis in original).

7. Ibid., 98.

8. Ibid., 92, 98.

9. Scharff, "Historical Material," 2.

10. Scharff's 1938 history of the refuge states, "The story of the Malheur is typical of the unfortunate history of waterfowl throughout the arid West, where uncontrolled shooting, over-grazing of the marsh areas, and the general abuse inflicted on this class of game, took an immense toll. The story of the wildlife abuses and the possibility of correcting these abuses was brought to the attention of . . . President Theodore Roosevelt with the result that on August 18, 1908 he signed a special executive order establishing Malheur, Harney, and Mud Lakes a combined area of 88,960 acres as a Federal wildlife reservation" (Scharff, "A Brief History of Malheur Refuge," 5).

11. Vileisis, *Discovering the Unknown Landscape,* 151. For a much fuller examination of the role of women in bird conservation, see Price, "When Women Were Women," 57–110. On the growing national concern over bird populations, see Barrow, *A Passion for Birds;* Gabrielson, *Wildlife Refuges;* and Dunlap, *Saving America's Wildlife.*

12. Reed and Drabelle, *The United States Fish and Wildlife Service,* 5. In 1871 the Bureau

of Fisheries was established and later incorporated into the Department of Commerce. In 1886 the Division of Economic Ornithology and Mammalogy was established in the Department of Agriculture and renamed the Division of the Biological Survey in 1891. These two agencies were the ancestors of the Fish and Wildlife Service.

13. Ibid., 7.

14. Gabrielson, *Wildlife Refuges*, 11.

15. Vileisis, *Discovering the Unknown Landscape*, 155.

16. Ibid.; Reed and Drabelle, *The United States Fish and Wildlife Service*, 9.

17. The administrator of the Blitzen properties began a drainage scheme in 1902. The papers of the French-Glenn Livestock Company detail the progress of this project: "Swamp reclamation: Near the center of the property is a large body of the richest kind of tule swamp land. It occupies in round numbers 10,000 acres at the north end of the 'P' Ranch Valley, an equal amount in the south end of Sod-House Division, the two separated from each other by a narrow gorge through which the river runs and about 5,000 acres in the lower end of the Diamond Division. This land is all owned by the Company. Many tule swamps have been reclaimed and invariably constitute the best meadows. Dredger on boat 75 feet 18 feet wide cost $15,000, digging canal 9 ft. deep 18 ft. wide on bottom and 24 ft. wide on top, Juniper wood used as fuel cuts 3,000 to 3,800 of ditch per month. Dredger started at point on river near southwest corner 26 Tp. 28 R. 31 E. ran southerly through Sec. 35 in that township and through Sections 2, 11, 14 and 23 in Tp. 29 R. 31 E. to the north end of the river gorge, 13 miles at present" (Blitzen Valley Land Company, "General Description of the Property," 2).

18. Ibid., 2, 3.

19. *East Oregon Herald,* July 18, 1903.

20. Waring, "Geology and Water Resources," 48.

21. Vileisis, *Discovering the Unknown Landscape,* 77.

22. Monroe, *Feelin' Fine!,* 173, 171.

23. Ibid., 175.

24. Whistler and Lewis, "Harney and Silver Creek Projects," 30.

25. Blitzen Valley Land Company, "General Description of the Property," 2–3, 4.

26. Vileisis, *Discovering the Unknown Landscape,* 84, 125. Between 1870 and 1900, twelve states passed legislation for drainage district formation, and over 14.5 million acres of wetlands came under their control.

27. State of Oregon, "Report of the Desert Land Board," 21.

28. Finley and Bohlman, "Is Malheur Lake Not Worth Saving?"

29. In the 1921 report of the Desert Land Board, eight segregations in Harney County were listed (State of Oregon, "Report of the Desert Land Board," 5th biennial report).

30. Lewis, "Irrigation in Oregon," 30.

31. Waring, "Geology and Water Resources," 52.

32. Ibid.

33. T. Gilbert Pearson, the secretary of the National Association of Audubon Societies, wrote in 1916 to Frank Triska, refuge warden, "I find that the State of Oregon has filed some sort of a claim there for a title to the lake" (Refuge files, Malheur National Wildlife Refuge, Princeton, Oregon).

34. "While the soils of this project are generally deep, of good texture and well supplied with plant food, they are too strong in soluble alkali salts in places. Drainage with proper irrigation affords the best means of alkali control. . . . Flushing off alkali in autumn from the salt grass areas may prove feasible where topography is fairly uniform and adequate outlets are close at hand" (Whistler and Lewis, "Harney and Silver Creek Projects," 26).

35. Michael Robinson, *Water for the West*, 46.

36. Whistler and Lewis, "Harney and Silver Creek Projects," 55, 63.

37. Ibid., 8.

38. Ibid., 39.

39. Ibid., 76.

40. A. J. Wiley, "Report 1922."

41. Gabrielson, *Wildlife Refuges*, 12. After the 1908 executive order, John Scharff wrote in 1972, "For a considerable number of years following the creation of the Malheur refuge only meager protection was afforded the waterfowl as no right to use the water was involved in the acquisition nor could any control be exercised over grazing or any other use be made of the unit" (Scharff, "Historical Material," 2).

42. Gabrielson, *Wildlife Refuges*, 4–5.

43. Finley, letter to Mr. McNary.

44. Finley and Bohlman, "Is Malheur Lake Not Worth Saving?"

45. Ibid.

46. Ibid.

47. Finley, letter to Mr. McNary.

48. Finley, "Malheur, The Unfortunate."

49. Willet, "Report on Malheur Lake Bird Reservation," 2, 6, 7.

50. "If the control of this lake can be secured by the Biological Survey, steps should be promptly taken to regulate, so far as is possible, the water level of the lake, this regulation being most important, both to the interests of the birds and the property owners around the lake" (ibid., 7).

51. Ibid., 8.

52. Ibid., 7. I cannot find the derogatory Forest Service report, but such a report would suggest several things. First, some engineers had thought this effort at water control was misguided—so there was no unanimity of opinion at the time. Second, Willet suspected that the Forest Service men could be corrupted by the drainage interests, but there is no way to figure out if his suspicion was well founded or if he was just a cynic.

53. Ibid., 9.

54. Letter to *Portland Oregonian*, July 28, 1930.

55. Donegan, untitled typed manuscript.

56. Hanley, "Opinions of People."

57. Quoted in Ibid.

58. Ibid. The firm quoted was Barr & Cunningham, Irrigation Engineers, Portland, Ore.

59. Ibid.

60. USDI Fish and Wildlife Service, "Compromise Committee Report," 3.

61. Ibid.

62. Willet, "Report on Malheur Lake Bird Reservation," 3.

63. Otley, letter to E. W. Nelson.

64. Ibid.

65. Fred Otley, "Malheur Lake Riparian Owners Plea," 4.

66. Otley, letter to E. W. Nelson.

67. Otley, "Malheur Lake Riparian Owners Plea," 5.

68. Otley, letter to E. W. Nelson.

69. Finley and Bohlman, "Is Malheur Lake Not Worth Saving?"

70. Liljiqvist, statement.

71. Finley and Bohlman, "Is Malheur Lake Not Worth Saving?"

72. Brimlow, *Harney County Oregon*, 247.

73. James Donegan, quoted in ibid.

74. *Crane American*, "Grain," August 4, 1934.

75. George Benson, letter to chief of Biological Survey.

76. Finley, letter to Mr. McNary.

77. Dunlap, USDA acting secretary, letter to Finley; Finley and Bohlman, "Is Malheur Lake Worth Saving?"

78. Blake et al., *Balancing Water*, 51.

79. Bureau of Reclamation website, http://dataweb.usbr.gov/html/klamath.html#plan.

80. Jessup, "Report."

81. The secretary of the interior authorized development of the project on May 15, 1905, under provisions of the Reclamation Act of 1902 (32 Stat. 388) (Bureau of Reclamation web site, http://dataweb.usbr.gov/html/klamath.html#plan).

82. Ibid.

83. Blake et al., *Balancing Water*, 77.

84. Amory, "Agriculture and Wild Fowl Conservation," 80–82.

85. Marbut report cited in ibid. and in Finley, "Reclamation in the Klamath District," 2.

86. Ibid.

87. Finley, "The Marshes of the Malheur." In 1923 the Reclamation Service's name changed to the Bureau of Reclamation.

88. Gabrielson, *Wildlife Refuges*, 13.

89. These letters are collected in the Oregon State University Archives, Finley Papers Box 7, Klamath-Reclamation folders.

90. Vileisis, *Discovering the Unknown Landscape,* 173–75.

91. Beck et al., "Report of the President's Commission," 4. The report's description of problems stemming from drainage are succinct and pithy: "The rapid depletion of the migratory waterfowl resource, now universally admitted to be a fact, is in large part a result of the unwise exploitation of sub-marginal lands. Drainage operations, intended to bring more land under cultivation, have directly destroyed millions of acres of former breeding grounds, and by lowering of water tables, have indirectly destroyed millions of acres more. Grazing of the remaining marshlands and ranges has prevented successful nesting and reproduction of breeding stocks. Mowing of hay and fires have destroyed many nests and nesting sites. This destruction of nests by grazing and mowing the shores of lakes and sloughs has reduced the annual increase from a normal expectancy of 300 percent to as low as 15 percent in areas under observation " (p. 11).

92. Ibid., 6.

93. In 1905 the Division of Biological Survey became the Bureau of Biological Survey, and in 1939 was transferred from the U.S. Department of Agriculture to the Department of the Interior. In 1940 the Bureau of Biological Survey and the Bureau of Fisheries merged to form the Fish and Wildlife Service.

94. Vileisis, *Discovering the Unknown Landscape,* 173–79.

95. Letter to Finley, May 15, 1934.

96. Letter to Finley, May 21, 1934. Darling expressed to Finley his frustration at his failure to create a national program to fund refuge acquisition: "I wish lots of things right now and one of the chief of them is at least some of the money promised by the Federal Government for the acquisition of migratory waterfowl resting areas would be forthcoming. So far we have had nothing but promises and not a cent has been allocated to this Department which we may spend even on investigations of the areas which we hope to buy. I know how mad you will be on getting this news and you are in no different emotional state than are we here in Washington, but we have hammered and pounded and tried in an effort to loosen up the guardian knot. Maybe we will finally succeed in getting a few crumbs but the original hope of a nationally planned program has almost died"(ibid.).

97. Letter to Darling, June 2, 1934.

98. Talbot, "P Ranch," 128–29.

99. "P-Ranch Bird Refuge Sold for $610,000," *Crane American,* Aug. 3, 1934, Refuge files, Malheur National Wildlife Refuge, Princeton, Oregon.

100. *Oregon Daily,* Feb. 21, 1935, Refuge files, Malheur National Wildlife Refuge, Princeton, Oregon. The article reads, "Sale of the historic P Ranch, 65,000 acre cattle barony in Harney county, famed in the romantic history of the Oregon cow country, for use as a federal game refuge was announced today by Carl C. Donaugh, United States district

attorney, who turned over a check for $675,000 to the Eastern Oregon Livestock company to complete the largest real estate transfer in the state this year.

"It was Peter French, pint-size sower of the seeds of empire, who fought the wilderness and Indians and battled hostile ranchers to carve out his vast holdings in Donner und Blitzen valley in the riotous, gun-fighting days of 1870, who established the P Ranch, which was the headquarters of his sprawling acres. . . . Here gathered such fabled great ones as 'Hen' Owen, John Devine, Bill Hanley, and others of the gallant crew who ran their teeming herds of bawling cow critters on the lush pasturage of Blitzen valley, Happy valley and Catlow valley."

101. *Portland Oregonian,* Feb. 23, 1935, Refuge files, Malheur National Wildlife Refuge, Princeton, Oregon.

4 / MANAGING DUCKS

1. Cameron and Ivey, interview. The "Annual Narratives" and "Quarterly Narratives" from 1935 to 1996, stored in the Malheur National Wildlife Refuge Station Library, provide a good record of changing management practices.

2. Quoted in Hays, *Conservation and the Gospel of Efficiency,* 124.

3. Beck et al., "Report of the President's Commission," 4.

4. Ibid., 9.

5. Gabrielson, *Wildlife Refuges,* 5.

6. Ibid., 34.

7. Jewett and Finley, "Malheur Lake."

8. Vileisis, *Discovering the Unknown Landscape,* 180.

9. Jewett, memo to Darling.

George Benson, refuge warden, wrote to Ding Darling on November 27, 1934, describing his ideas for restoration, including restoration of beaver and sage hen populations, planting of willows, and extensive water control. Among his proposals were "No. 1. Remove all unnecessary dams in Blitzen river. Reconstruct those beneficial in water control. . . . No. 3. Divide the reservation into units by constructing dams and levies with head gates control. No. 4. Turn the Blitzen river channel within the Springer ranch north by dredging or scraping out an old slough going north and making connection with the lilly slough which enters the lake about one half mile west of lake gauge no 1. By changing the river as mentioned it would increase a large slough area suitable for a nesting bird. The location would be suitable for tree planting and other shrubbery. This change would leave the big spring and channel to be developed in a separate unit, by installing lot dams to hold the water at a reasonable depth, fish could be planted. Trees and shrubbery planted on the banks for shade and wind break. It would also prove beneficial in holding crippled birds and many other uses of attraction. . . . No. 11. I recommend strict supervising of the Blitzen river valley and the diamond swamp area known as the P ranch in order to protect the sage hen which is fast increasing, also protect the beaver colonies along the Blitzen river. Roads should be constructed through central part of valley for patrol work only, good highways should be constructed on each side of valley to afford travel a good view of wild life. A dam should be constructed at what is known the diamond crossing in the narrow part of the valley in order to re-flood the swamp area below the P ranch house. This dam would only need to be about 10 ft. high in order to flood a large area to bring back thousands of nesting birds" (Benson, letter to Ding Darling).

10. Jewett, memo to Ding Darling.

11. Griffin, *Starting at the Narrows,* vii.

12. Jacoby, *Crimes Against Nature;* Warren, *The Hunter's Game.*

13. Finley, "Malheur Refuge an Aid to Both Birds and Farmers."

14. Finley felt the federal government was not doing much to protect the refuge, writing that "the leniency of the Biological Survey in dealing with some of the people who have squatted on the lake bed has brought on much more trouble than if this federal property had been rigidly protected. No squatters now living within the meander line of the lake have ever been given titles to the land by the government. They have fenced and are using government property. They pay no county or state taxes, and therefore, have an unfair advantage over bonafide residents around the lake. Recognizing those factors, the Biological Survey is now taking steps to fence the entire lake bed" (ibid.).

15. Wallace, letter to Hon. Walter M. Pierce.

16. Scharff, "Report of Activities Fiscal Year 1939," 4.

17. Simpson, *The Community of Cattlemen,* 153.

18. Finley, "Malheur Refuge an Aid to Both Birds and Farmers."

19. Finley, "Malheur Lake."

20. Finley, "The Marshes of the Malheur."

21. Sharp, *Where Rolls the Oregon,* 58.

22. Jewett and Finley, "Malheur Lake—Blitzen Valley Migratory Waterfowl Refuge."

23. Scharff, "Historical Material—P-Ranch," 1.

24. Ibid.

25. Ibid.

26. Scharff, "Report of Activities Fiscal Year 1937," 12.

27. Ibid., 5.

28. Ibid., 15.

29. Ibid., 5, 15.

30. Scharff, "A Brief History of Malheur Refuge."

31. H. S. Davis, address to 1934 annual meeting of the American Fisheries Society, quoted in Hunter, *Better Trout Habitat,* 7.

32. Talbot, "P Ranch": "Every spring, ranch crews reinforced the rock irrigation dams near the long barn with manure and hay; one is still visible on the river just to the east and upstream. Scharff relates that engineer Page effectively riprapped banks along the

damsites with layers of willow and sagebrush. High water occurred around the barn in spring, and the ditch running from the Blitzen past the barn was the main outlet for water in the area" (48). "Traditionally, the sudden spring breakup made it expedient that for several nights until the meltwaters peaked, a crew of ranch hands had to keep a vigil at the low bridge, pushing ice blocks through with long-handled pokers to prevent a buildup of debris and the subsequent destructive flooding of much of the ranch area" (61).

33. Ibid., 44.

34. USDI Fish and Wildlife Service, Malheur National Wildlife Refuge, *Annual Narrative 1946*, 7.

35. Finley, "Reclamation vs. Conservation," 46–48.

36. Finley, "Trapping and Transplanting Beaver," unpaginated.

37. Ibid.

38. Finley, "Habits and History of the Beavers," unpaginated.

39. Jewett, memo to Ding Darling, 6.

40. Finley, "Malheur Lake–Blitzen Valley Migratory Waterfowl Refuge."

41. Markey and Cramer, *Beaver Survey*, Markey notes for August 9, 1937; Cramer notes for August 12, 1937.

42. Markey and Cramer; Markey notes for August 9, 1937; Cramer notes for August 14 and 12, 1937.

43. Markey and Cramer; Cramer notes for August 14, 1937: "Proceeded to Malheur Lake Refuge where Mr. Springer and rancher near the headquarters, had made a complaint. Talked with Mr. John Scharff, manager of the refuge. He said he hoped to plant a few beaver on Paul Stewart's ranch. He also planned to plant some beaver from the Diamond Valley country to the refuge in the vicinity of the P Ranch."

44. USDI Fish and Wildlife Service, Malheur National Wildlife Refuge, *Annual Narrative 1947*, 8.

45. USDI Fish and Wildlife Service, Malheur National Wildlife Refuge, *Quarterly Narrative Report January–April 1946*, 6.

46. USDI Fish and Wildlife Service, Malheur National Wildlife Refuge, *Quarterly Narrative Report January–April 1947*.

47. USDI Fish and Wildlife Service, Malheur National Wildlife Refuge, *Annual Narratives*, 1947, 1950, 1954, and 1955.

48. In 1944 Congress allotted substantial funds for Soil Conservation Service's drainage and irrigation work, and memos clarified that drainage and irrigation were "technical conservation practices" (Vileisis, *Discovering the Unknown Landscape*, 196).

49. Ibid., 197.

50. Ibid., 201. The Fish and Wildlife Service and the Soil Conservation Service adopted a Memorandum of Understanding, in which federal and state wildlife agency recommendations were supposed to be incorporated into watershed plans. Local soil conservation districts had final say, however, and channelization soon began in earnest.

51. Vileisis, *Discovering the Unknown Landscape*, 245.

52. USDA Economic Research Service, "Report on Water," 119, 83, 91, vii, 102.

53. Vileisis, *Discovering the Unknown Landscape*, 249, 251.

54. Jewett wrote "steps should be taken immediately to plant willows, particularly on Cole Island and Cane Island, and investigation was made as to the practicability of putting them on other parts of the reservation. The warden in charge reported that certain individuals were in the habit of burning the tules on Cole Island each year, this part of the reservation being used as winter pasture and feed grounds for a number of cattle. This practice should be discontinued, especially if willows are to be planted to increase facilities for nesting birds" (Jewett, "Report of Malheur Lake Bird Reservation," 2).

55. Scharff, "Report of Activities," fiscal years 1937 (10), 1938 (13), and 1939 (10).

56. Scharff. "Report of Activities Fiscal Year 1937," 13.

57. USDI Fish and Wildlife Service, Malheur National Wildlife Refuge, *Quarterly Narrative Report,* January–April 1943, May–August 1943, May–August 1949.

58. USDI Fish and Wildlife Service, Malheur National Wildlife Refuge, *Quarterly Narrative Report September to December 1955,* 9–10.

59. Ibid., 10.

60. USDI Fish and Wildlife Service, Malheur National Wildlife Refuge, *Quarterly Narrative Report September to December 1956,* 10; USDI Fish and Wildlife Service, Malheur National Wildlife Refuge, *Quarterly Narrative Report September to December 1957,* 8.

61. USDA Economic Research Service, "Report on Water," 87.

62. Fletcher and Elmendorf, "Phreatophytes—A Serious Problem in the West," 424.

63. On spraying of 2,4-D and 2,4,5-T, see ibid., 427. One of the first studies to draw scientific attention to the "waste" posed by phreatophytes was O. E. Meinzer, "Plants as Indicators of Ground Water," U. S. Geological Survey Water Supply Paper 577 (1927). See the extensive review of the scientific literature in Robinson, "Phreatophytes." In 1970 Forest Service botanist C. J. Campbell urged managers to consider the possible ecological importance of phreatophytes, but his caution was largely ignored for two decades (Campell, "Ecological Implications of Riparian Vegetation Management").

64. Cattron, letter to Frank Triska.

65. Cantwell, letter to Frank Triska.

66. Talmer, letter to George Cantwell.

67. Willet, "Report of Malheur Lake Bird Reservation," 5.

68. Stone, letter to George Benson.

69. Sheldon, letter to George Benson.

70. Deumead (?), letter to George Benson.

71. Benson, letter to Chief of Biological Survey. "I believe this is a very important move to keep live stock out of the nesting area, especially in the nesting season," Benson wrote. "But I realize under the present condition it is going to be a rather difficulty [sic] proposition to handle. The east side of the reservation is open to the outside range, and

stock . . . depend more or less on Malheur lake for drinking water, especially the last 4 or 5 years, during this dry period most of the springs in the foot hills have gone dry, and this condition has compelled the range stock to go to the lake for water, under this condition is when the stock does most damage to the nesting areas. . . . It is a desperate condition where there is a shortage of water for live stock, and the proposition will need carefully [sic] consideration."

72. Jewett and Leichhardt, letter to Randolph S. Collins.

73. Moran, "Memo," 4.

74. Finley, letter to Harold C. Ickes.

75. Finley, letter to H. M. Worcester. After the Hart Mountain Game Range was created, Finley wrote to Ickes in a fury, for he felt the executive order wrongly allowed far too much grazing and far too little habitat for antelope: "The Hart Mt. Game Range has been wrongly named. The wording of the Executive Order makes this a Hart Mt. Stock Range. While this is the greatest antelope country now existing in the U.S., the order cuts our game species to the very minimum, far below what this range could support without in any way interfering with stock interests and giving stock the primary preference in a wide range where water holes are very limited.

"It was my understanding that your ideas of conservation were not to turn all the public property that we have over to private interests. It was also my understanding that you have the authority to set aside a few small portions to be kept under public control where our disappearing big game species of the West can have at least a few water holes and guaranteed grazing rights" (Finley, letter to Harold Ickes).

76. Finley and Bohlman, "Is Malheur Lake Not Worth Saving?" 2.

77. Scharff, "Report of Activities Fiscal Year 1937."

78. Scharff, "Report of Activities Fiscal Year 1938," 16.

79. USDI Fish and Wildlife Service, Malheur National Wildlife Refuge, *Quarterly Narrative Report August–October 1939.*

80. Scharff, "Report of Activities Fiscal Year 1938," 41.

81. USDI Fish and Wildlife Service, Malheur National Wildlife Refuge, *Quarterly Narrative Report January–April 1947.*

82. USDI Fish and Wildlife Service, Malheur National Wildlife Refuge, *Quarterly Narrative Report January–April 1947,* 17.

83. In Idaho, carp were introduced in the 1880s and 1890s as a food source (Fiege, *Irrigated Eden,* 57). Refuge biologist Harold Duebbert wrote that carp were introduced into the Silvies River watershed in the 1920s, but gives no source (Duebbert, "The Ecology of Malheur Lake").

84. USDI Fish and Wildlife Service, Malheur National Wildlife Refuge, *Quarterly Narrative Report September to December 1955,* 2.

85. Ibid.

86. Ibid., 11–12.

87. USDI Fish and Wildlife Service, Bureau of Sport Fisheries and Wildlife, "Carp Control Project," 15.

88. USDI Fish and Wildlife Service, Malheur National Wildlife Refuge, *Quarterly Narrative Report September to December 1955*, 12.

89. USDI Fish and Wildlife Service, Bureau of Sport Fisheries and Wildlife, "Carp Control Project," 35.

90. USDI Fish and Wildlife Service, Malheur National Wildlife Refuge, *Annual Narrative 1977*, 9; USDI Fish and Wildlife Service, Malheur National Wildlife Refuge, *Annual Narrative 1987*, 19.

91. USDI Fish and Wildlife Service, Bureau of Sport Fisheries and Wildlife, "Carp Control Project," 15.

92. TABLE 4.1: Native Fish Species, 1936

Common Name	Scientific Name	Location
rainbow trout	*Salmo gairdneri*	Blitzen R., Boca L.
mountain whitefish	*Prosopium williamsoni*	Blitzen R.
coarse-scaled sucker	*Catostomus macrochelius*	Lower Silvies R.
chiselmouth	*Acrocheilus alutaceus*	Silvies R.
longnose dace	*Rhinichthys cataractae*	Blitzen R.
freshwater sculpin	*Cottus* sp.	Blitzen R.
fine-scaled sucker	*Catostomus syncheilus*	Blitzen R., Silvies R.
roach	*Siphateles bicolor*	Boca L., Diamond Cr.
squawfish	*Ptychocheilus oregonense*	Blitzen R., Silvies R.
redside shiner	*Richardsonium balteatus*	Blitzen R. Silvies R.

SOURCE: USDI Fish and Wildlife Service, Bureau of Sport Fisheries and Wildlife, "Carp Control Project," 9.

93. Ibid., 36.

94. Ibid., 42.

95. Malheur Refuge might have considered what Dane County, Wisconsin, did when faced with its own version of carp abundance: officials turned the carp into a cash crop by netting them out of Lake Kegonsa at Fish Camp (now a park) and tanking them out live on the railroad to sell in Chicago and New York City. Although the Wisconsin project did not ultimately reduce the number of carp in the lakes, it made something useful out of the carp: "The Kegonsa Rough Fish Station [was] run by the old Wisconsin Conservation Department for almost 20 years beginning in the late 1940's. . . . Using four barges and about 4000 feet of seine, employees netted unbelievable amounts of carp, sheepshead and buffalo fish from the four lakes. The fish were kept in fenced 'ponds' next to the sta-

tion, and most were tanked out live by rail to Chicago and New York City" ("Bike Routes," Town of Dunn, Wisconsin, Park Commission brochure, 1983).

96. Brimlow, *Harney County Oregon,* 248.

5 / GRAZING, FLOODS, AND FISH

1. Ferguson and Ferguson, *Sacred Cows at the Public Trough,* 113.
2. Ibid., 1–2.
3. Ibid., 1–2, 109–12. Refuge biologists argue that carp, not cows, are likely to have been the major cause for waterfowl declines (Roy, interview, January 10, 2002; see also Ivey et al., "Carp Impacts on Waterfowl").
4. Walters, *Surviving the Second Civil War,* 4ff.
5. Cramer, letter to *Burns Times Herald.*
6. Mazzoni, *Annual Narrative 1974,* 14.
7. Mazzoni, *Annual Narrative 1975,* 13.
8. Idem, *Annual Narrative 1977,* 10.
9. Ohmart, "Historical and Present Impacts of Livestock Grazing," 255.
10. Elmore and Beschta, "Riparian Areas," 261.
11. Chris Sokol, "Rivers from a Timber Industry Perspective," 37.
12. Durbin, "Restoring a Refuge."
13. Southworth, "Ranching and Riparians," 12–13.
14. Mazzoni, *Annual Narrative 1977,* 8.
15. USDI Fish and Wildlife Service Malheur National Wildlife Refuge *Annual Narrative 1987,* 34.
16. George Constantino, *Annual Narrative 1987,* 20.
17. Durbin, "Ranchers Arrested at Wildlife Refuge."
18. Ibid.
19. Cameron, interview with author.
20. Jackman and Scharff, *Steens Mountain,* 16
21. Approved August 18, 1941, Public Law No. 288, 77th Congress, 1st Session: "The Secretary of War is hereby authorized and directed to cause preliminary examination and surveys for flood control, to be made under the direction of the Chief of Engineers, in drainage areas of the United States and its territorial possessions " (quoted in U.S. Army Engineer District, Portland, Corps of Engineers, "Survey Report," 1).
22. Ibid., 6, 12.
23. Ibid., 29.
24. Ibid., 15.
25. Ibid., 29.
26. Ibid., 40, 28, 50.
27. Ibid., 56, 60, 66.

28. In 1964 a flood had caused $2,486,000 worth of damage in the basin, the most expensive yet. Local feelings, therefore, were strong about the need for flood control, at least during early public hearings in 1969. However, the army noted that by the 1976 public hearing, ranchers were presenting concerns about the usual issues: the cost of stored water, the acreage limitation imposed by a federal project, and the potential changes in ranching practices (U. S. Army Corps of Engineers, "Silvies River and Tributaries," 10).

29. For costs calculations, see ibid., 34. The high annual costs in the 1977 estimates were driven partly by the higher initial charges for the project construction, as well as annual payments on the $1,150–per-acre costs of draining land and creating an irrigation distribution system (converted over the expected hundred-year life of the project at 6.12 percent interest).

30. Ibid., syllabus.

31. U.S. Army Engineer District, "Survey Report," 32.

32. U.S. Army Corps of Engineers, "Silvies River and Tributaries," 11.

33. Ibid., 32.

34. Raleigh, letter to Col. Allaire, 17.

35. Stivers, letter to Col. C. J. Allaire.

36. Raleigh, letter to Col. Allaire, 19.

37. Findlay, letter to Col. Robert Giesen; Findlay, letter to district engineer; Geren, letter to Col. Richard M. Connell.

38. Since records have been kept, the normal lake level maximum has been 4,093 feet above sea level. Of course, normal lake levels do not mean a great deal, for in a closed basin, the level varies each year depending on snowfall and patterns of snowmelt (Mandaville, "A Swamp in the Desert," 64).

39. U.S. Army Corps of Engineers, "Final Version," 2–4.

40. Arp, "The Flooding of Malheur Lake," 6; Braymen, Burns Times Herald.

41. Arp, "The Flooding of Malheur Lake," 11.

42. Mandaville, "A Swamp in the Desert," 65.

43. U.S Army Corps of Engineers, "Final Version," summary. For example, in 1985 assessed valuation of property in Harney County dropped $12,280,000, most of it from the floods (Arp, "The Flooding of Malheur Lake," 6). According to Bill Beal, county water master, arsenic levels increased; for example, Malheur Field Station had to drill a new well because arsenic levels went up from 0.04 to 0.12 (unit of measurement not given) (ibid., 8). Meanwhile, carp and other undesirable fish increased to 80 to 90 percent of the fish biomass in the lake (ibid., 2–12).

44. Arp, "The Flooding of Malheur Lake,"6.

45. U.S. Army Corps of Engineers, "Reconnaissance Report," 17.

46. Ibid., 33.

47. U.S. Army Corps, "Final Version," 3–13.

48. U.S. Army Corps, "Reconnaissance Report," 7. According to the Squaw Butte's

own inflated estimates of hay value in 1977, meadow hay was worth $20 per ton, and meadows without irrigation could produce .75 tons per acre, so that fifty thousand acres could produce $750,000 worth. The highest estimate for the value of hay lost over the entire flooding was $1.2 million.

49. The proposed canal was seventeen miles long, with a fish barrier to "minimize movement of carp and Tui chub from Malheur Lake into the Malheur River system" (U.S. Army Corps, "Final Version," 3–9).

50. U.S. Army Corps, "Final Version," summary.

51. Ibid., 2–9. Malheur Lake water quality was usually better than that of Harney and Mud Lakes because during periods of high water, Malheur normally would flush into the latter two, slightly lower-lying lakes. Concentrations of most dissolved solids were normally ten times higher in Harney Lake than in Malheur, but the very high lake levels meant some mixing of the lakes had occurred, and this poorer quality water would have been flushed into Malheur River if the canal had been completed.

52. Ibid., 2–11. Concentrations of calcium, magnesium, sodium, chloride, arsenic, and boron were higher downstream, and increased later in the irrigation season.

53. Ibid., 2–12. Below Namorf were unproductive warm water fisheries; the Army Corps report stated that water withdrawal for irrigation—which led to low flows, warm temperatures, heavy silt loads, and overall poor habitat—was the primary cause of the poor quality.

54. Ibid., 2–13. Until 1973 the Oregon Department of Fish and Wildlife had periodically used rotenone to chemically eradicate native nongame fish species and replace them with rainbow trout. But for several reasons, no poisoning or restocking had occurred between 1973 and 1987, when the project report was written: uncertainty about flooding effects and potential construction projects; high rotenone prices, which had quintupled during those years, from $7 to $35 a gallon; focusing of Department of Fish and Wildlife attention on native species such as Eagle Lake rainbow and Lahontan cutthroat trout; and a shift in departmental management strategy toward developing more wild stock and less hatchery stock.

55. USDI Fish and Wildlife Service, Division of Ecological Services, "Planning Aid Letter," C-13.

56. U.S. Army Corps, "Reconnaissance Report," 11, 12.

57. USDI Fish and Wildlife Service, Division of Ecological Services, "Planning Aid Letter," C-12.

58. Ibid., C-3, C-12.

59. Ibid.

60. USDI Fish and Wildlife Service, Malheur National Wildlife Refuge, *Annual Narrative 1986*.

61. Duebbert, "The Ecology of Malheur Lake," 20.

62. Ibid. Scharff's plan proposed such a water control system (USDI Fish and Wildlife Service, Bureau of Sport Fisheries and Wildlife, *Master plan*).

63. Ibid., 21. Despite all the manipulations at Malheur, it was still one of the most natural marshes in the entire refuge system.

64. U.S. Army Corps of Engineers, "Final Version," summary.

65. Ibid., 1–3.

66. Malheur Refuge web site, http://www.rl.fws.gove/malheur/prehfish.htm/.

67. Gerald Smith, University of Michigan, cited in Oregon Natural Desert Association, Oregon Trout, Native Fish Society, and Oregon Council of Trout Unlimited, "A Petition," 18.

68. Cameron, interview.

69. Oregon Natural Desert Association et al., "A Petition," 5, citing USDA/USDI "Interior Columbia Basin Ecosystem Management Project"; Kostow, "Biennial Report."

70. Oregon Natural Desert Association et al., "A Petition," 6, 15; Howell, "Oregon Chapter Workshop," 34–36.

71. Oregon Natural Desert Association et al., "A Petition,"2.

72. Ibid., 7, citing Behnke, *Native Trout;* and Kostow, "Biennial Report."

73. Oregon Natural Desert Association et al., "A Petition," 8.

74. Ibid., 14; Behnke, *Native Trout.*

75. Oregon Natural Desert Association et al., "A Petition," 6.

76. Ibid., 9–10.

77. Ibid., 10.

78. Ibid., 10, citing Behnke, *Native Trout.*

79. Oregon Natural Desert Association et al., "A Petition," 11, citing Lee et al., "Broadscale Assessment of Aquatic Species and Habitats."

80. Hunter, *Better Trout Habitat,* 39, 53.

81. Gregory et al., "An Ecosystem Perspective of Riparian Zones," 548.

82. USDA Forest Service, *Malheur National Forest Land and Resource Management Plan,* 111–35.

83. Gregory et al., "An Ecosystem Perspective of Riparian Zones," 548.

84. Li and Lambaerti, "Cumulative Effects of Riparian Disturbances," 627–40; Lee et al., "Broadscale Assessment of Aquatic Species and Habitats"; Oregon Natural Desert Association et al., "A Petition," 20; Bowers et al., "Conservation Status of Oregon Basin Redband Trout."

85. Lee et al., "Broadscale Assessment of Aquatic Species and Habitats."

86. Burns BLM water temperature data, in Oregon Natural Desert Association et al. "A Petition," fig. 9, p. 20.

87. Lee et al., "Broadscale Assessment of Aquatic Species and Habitats"; Kostow, "Biennial Report."

88. On temperatures in pools, see: Oregon Natural Desert Association et al., "A Petition," 21. On irrigation effects, see Lee et al., "Broadscale Assessment of Aquatic Species and Habitats." On the Lahontan study, see Dunham et al., "Habitat Fragmentation."

89. On effects of dams on genetic diversity, see Oregon Natural Desert Association et al., "A Petition," 23. On adfluvial forms of redband trout, see Kostow, "Biennial Report."

90. Bowers et al., "Conservation Status of Oregon Basin Redband Trout," 2.

91. Ibid., 5; citing unpublished data from refuge staff.

92. Ibid., 6; Hosford and Pribyl, "Blitzen River Redband Trout Evaluation."

93. Ibid., 51. It is important to note that hatchery introductions have not affected Great Basin redband trout as much as they have harmed many other species, largely because the hatcheries introduced rainbow trout from the west side that could not survive long in the desert waters.

94. Richard Roy (supervisory wildlife biologist, Malheur National Wildlife Refuge), "Blitzen River Fish Habitat Improvement Strategy"; interview with Roy, January 10 and 11, 2002.

95. USDI Fish and Wildlife Service, "Draft Environmental Impact," vol. 2, appendix O, 22–24, 38, 114, 118, 122.

96. Oregon Natural Desert Association et al., "A Petition,"36.

97. The initiative would have required that so-called "water quality limited" streams be protected from grazing impacts. These were streams that were determined to be in poor quality by Department of Environmental Quality criteria: temperature, taste, color, turbidity, silt, odor, fecal coliform, algae, aquatic weeds, and sediments. At least one-third of Oregon's streams had been assessed as water quality limited, and hence in need of improvement, or at least in need of protection from further degradation (Davis, "Polluted Waters Divide Oregon").

98. Wilson, "'Clean Stream Initiative.'"

99. Haggerty, "Opinion and Order," 6.

100. USDI Bureau of Land Management, "South Steens Allotment Management Plan, 3–4.

101. For example, one site in the allotment area had been described by David Griffith in 1902 as providing "no more feed than on the floor of a coral." In 1959 the Bureau of Land Management had set grazing levels at 21,926 AUMS on the public land in the allotment, with no restrictions on season of use, and no fencing between public and private lands. The 1982 BLM draft environmental impact statement for the Andrews area, which covered livestock grazing impacts in the watershed, made a series of predictions as to how grazing management would improve redband trout habitat. But the bureau was wrong: surveys in the 1990s showed that decline had in fact continued. The environmental impact statement had argued that fish habitat would improve on Home and Threemile Creeks, but in 1996 the redband population on those creeks was in danger of extinction, and habitat had not improved. For Skull Creek, the bureau had stated that proposed graz-

ing systems would result in poor conditions for fish. Ironically, this was the only correct prediction in the statement: fifteen years later, those fish were severely in danger of extinction (Oregon Natural Desert Association et al., "A Petition," 35). By the 1980s, as the bureau reported, "Monitoring studies indicated heavy to severe utilization in riparian zones accessible to livestock, with only slight to light use on most of the uplands." Even though most of the riparian areas were in below-satisfactory condition, the light use of the uplands meant that the bureau increased the allowable AUMS to 41,150 in 1989, just a year after the area was designated a Wild and Scenic River (USDI Bureau of Land Management, "South Steens Allotment Management Plan," 21–23).

102. Ibid., 43. Even in 1995, grazing pressures were most intense in riparian areas, with little use of most of the uplands. Although plan alternatives included both excluding grazing in the entire allotment and fencing off riparian areas, the preferred alternative continued to be riparian grazing. Under ideal conditions, with plenty of water tanks in the uplands to provide for cattle's needs, and with extremely devoted, committed riders, riparian areas would have improved on much of the allotment—but slowly.

103. Haggerty, "Opinion and Order," 3, 22, 29, 8; Hal Bernton, "Judge Tightens Grazing Rules"; USDI Bureau of Land Management, "Water Quality Management Plan," 31. As part of that agreement, the Bureau of Land Management agreed to build a fence that would protect part of the Blitzen River, creating the Blitzen Pasture, separating the south fork of the Blitzen and the uppermost reaches of Home Creek from the rest of the South Steens Pasture. The Blitzen Pasture was rested in 1997, and would continue to be so until the bureau completed an environmental impact statement, as ordered by the court, that analyzed the effects of grazing. The South Steens Pasture was rested in 1997 and grazed in 1998. This conservation agreement came about only because Judge Haggerty ruled that the bureau had violated the Wild and Scenic Rivers Act and NEPA. But, as ONDA pointed out, conservation agreements did not afford legal protection, for they were nonbinding (Oregon Natural Desert Association et al., "A Petition," 34).

104. In June 1998, as part of this agreement, the bureau prepared a water quality management plan, which continued to place most of the blame for riparian degradation and redband trout habitat degradation on juniper expansion, not on cattle: "Failure to meet the temperature standard and other water quality parameters is attributable to the condition of some areas of riparian habitat and areas of upland vegetation, especially where juniper encroachment is occurring." The plan insisted that grazing problems came from "past livestock grazing," not from current grazing (USDI Bureau of Land Management, "Water Quality Management Plan," 22).

105. USDI Fish and Wildlife Service, "Notice of Petition Finding." The Fish and Wildlife Service based its decision on historical analysis done by Oregon Department of Fish and Wildlife biologist J. Dambacher, who examined estimates of redband trout density from 1968 to 1995. Dambacher calculated that within those years, a low density was less than 0.059 fish per meter square, moderate density was between 0.06 and 0.19 fish per meter

square; and high density was over 0.2 fish per meter square. Since Dambacher's 1999 surveys showed redband densities to be 0.423 in Catlow, 0.372 in Harney, 0.216 in Warner Basin, 0.171 in Fort Rock basin, 0.143 fish in Chewaucan basin, and 0.140 fish in the Goose Lake basin, the Fish and Wildlife Service concluded that redband densities were moderate to high in all the basins, so therefore in no need of protection. This sounds good, but it's a difficult claim to evaluate, because redband populations from the period from 1968 to 1995 were already extremely low. Therefore, recovery to population levels that were moderate and high during those years of depressed populations doesn't necessarily mean recovery to a resilient population level.

106. Ibid.

107. Ibid., 14936.

108. Oregon Natural Desert Association, "Environmental Groups Disappointed."

109. Oregon Department of Fish and Wildlife, "Work must continue."

110. Ibid. The Department of Fish and Wildlife's press release stated that "many private landowners cooperated by limiting streamside grazing, restoring riparian vegetation, removing barriers to fish passage and installing screens to prevent fish from swimming into irrigation canals. Land management agencies modified both grazing and timber harvest practices to help restore streamside vegetation and remove fish passage barriers."

111. Stuebner, "Go Tell It on the Mountain."

112. Ibid.

113. Fred Otley, interview, June 1997. Richard Roy, interview, January 10, 2002.

114. Stuebner, "Go Tell It on the Mountain."

115. Ibid.

116. Ibid.

117. *Portland Oregonian*, "The Steens Legislation." See also *Portland Oregonian*, "The Fine Print on the Steens"; and Kerr, "Steens Mountain Wilderness Act of 2000."

6 / PRAGMATIC ADAPTIVE MANAGEMENT

1. *Portland Oregonian*, "Antelope Refuge Will Look at Goals."

2. Ibid.

3. Elaine Budlong, "Deja Vu."

4. Ibid.

5. *Portland Oregonian*, "Antelope Refuge Will Look at Goals."

6. James, "Pragmatism: A New Name," 28.

7. Ibid., 32.

8. Parker, "Pragmatism and Environmental Thought," 21, 23.

9. Dewey, "Pragmatic America," 59.

10. "John Dewey," in *The Oxford Companion to Philosophy*.

11. Parker, "Pragmatism and Environmental Thought," 25, 21.

12. Ibid., 30.

13. Rosenthal and Buchholz, "How Pragmatism *Is* an Environmental Ethic," citing James, "On a Certain Blindness in Human Beings." See also Fuller, "American Pragmatism Reconsidered."

14. James, "Pragmatism," in *The Works of William James,* 71.

15. Ibid.

16. Parker, "Pragmatism and Environmental Thought," 23; citing James, "A World of Pure Experience," 21–44.

17. James, "Pragmatism: A New Name," 53.

18. Ibid., 36.

19. "John Dewey," entry in *The Oxford Companion to Philosophy.*

20. Scott, *Seeing Like a State,* 4, 6.

21. Ibid., 342, 352, 43, 328, 330, 345.

22. Ibid., 342.

23. Shutkin, *The Land That Could Be,* xiv. See also Gottlieb, *Forcing the Spring,* 26ff.

24. Hickman, ed. *Reading Dewey,* xvi.

25. Parker, "Pragmatism and Environmental Thought," 27.

26. Rorty, "The Priority of Democracy to Philosophy," 196; idem, "On Ethnocentrism," 208; Guignon and Hiley, "Biting the Bullet," 339–40; Rorty, *Consequences of Pragmatism,* 166.

27. Dewey, "Liberalism and Social Action," 105.

28. Campbell, "Dewey's Understanding of Community," 27.

29. Ibid., 29.

30. Thompson, "Pragmatism and Policy," 190, 191.

31. Ibid., 200.

32. Ibid., 200, 205.

33. Gary Snyder, cited in Shutkin, *The Land That Could Be,* 140.

34. Mark Fiege, personal communication.

35. Worster, "A Country without Secrets," 252–53.

36. Young, James, and Jerry D. Buddy, "Historical Use of Nevada's Pinyon-Juniper Woodlands."

SELECTED BIBLIOGRAPHY

ABBREVIATIONS

RF   Refuge files, Malheur National Wildlife Refuge, Princeton, Ore.
USACE   U. S. Army Corps of Engineers, Walla Walla District
WFP   William Finley Papers, Oregon State University Archives

Amory, Copley. "Agriculture and Wild Fowl Conservation at Lower Klamath Lake." *New Reclamation Era,* May 1926, 80–82.

Arp, Gayland T. "The Flooding of Malheur Lake: The Problems and Solutions." Unpublished manuscript, Dec. 9, 1986. RF.

Barrow, Mark V., Jr. *A Passion for Birds: American Ornithology after Audubon.* Princeton: Princeton University Press, 1998.

Bastasch, Rick. *Waters of Oregon: A Source Book on Oregon's Water and Water Management.* Corvallis: Oregon State University Press, 1998.

Beck, Thomas H., Jay "Ding" Darling, and Aldo Leopold. "Report of the President's Commission on Wildlife Restoration," Feb. 8, 1934. WFP, box 9, Klamath/Malheur Materials, 1930s.

Beckham, Stephen Dow. "Donner und Blitzen River, Oregon: River Widths, Vegetative Environment, and Conditions Shaping Its Condition, Malheur Lake to Headwaters." Unpublished paper submitted to Interior Columbia Basin Ecosystem Management Project, Walla Walla, Wash., April 17, 1995. Available at <http://ww.icbemp.gov/science/scirpte.html>.

Behnke, R. J. *Native Trout of Western North America.* American Fisheries Society Monograph 6. Bethesda, M.D.: American Fisheries Society, 1992.

Benson, George. Letter to the chief of the Biological Survey. April 30, 1934. RF.

———. Letter to Ding Darling. Nov. 27, 1934. RF.

Bernton, Hal. "Judge Tightens Grazing Rules." *Portland Oregonian,* Feb. 4, 1997.

Blitzen Valley Land Company. "General Description of the Property Owned by the Blitzen Valley Land Company formerly known as the French-Glenn Property in Harney County, Oregon." Feb. 25, 1913. Company prospectus. Hanley Company Papers, Oregon Historical Society, ms. 378, box 4.

Bowers, Wayne, Roger Smith, Rhine Messmer, Curtis Edwards, and Ray Perkins. "Conservation Status of Oregon Basin Redband Trout, Public Review Draft." April 12, 1999. Oregon Department of Fish and Wildlife, Corvallis, Ore.

Braymen, Pauline. *Burns Times Herald,* Nov. 29, 1986.

Brimlow, George Francis. *Harney County Oregon, and Its Range Land.* Portland, Ore.: Binfords & Mort, 1951.

Budlong, Elaine. "Deja Vu: The Coyote as Fall Guy, One More Time." *Predator Project Newsletter,* Winter 1996.

Burnside, Carla (refuge archeologist). with author, June 23, 1997.

Burns Times-Herald. August 23, 1902; July 13, 1901. In Harney County Library, Burns, Ore.

Cameron, Forrest (project director, Malheur National Wildlife Refuge), and Gary Ivey (wildlife biologist, Malheur National Wildlife Refuge). Interview with author. July 25, 1996.

Campbell, C. J. "Ecological Implications of Riparian Vegetation Management." *Journal of Soil and Water Conservation* 25 (1970): 49–52.

Campbell, James. "Dewey's Understanding of Community." In *Reading Dewey: Interpretations for a Postmodern Generation,* ed. Larry Hickman, 23–42. Indianapolis: Indiana University Press, 1998.

Cantwell, George G. Letter to Frank Triska. Jan. 16, 1916. RF.

Cattron, E. S. (district inspector). Letter to Frank Triska. July 17, 1915. RF.

Constantino, George. *Annual Narrative 1987.* RF.

Couture, Marilyn Dunlap. "Recent and Contemporary Foraging Practices of the Harney Valley Paiute." Master's thesis, Portland State University, 1978. On file in Burns Paiute tribal office, Burns, Ore.

Cramer, William D. Letter to *Burns Times Herald.* Aug. 18, 1978.

Crane American. Aug. 3 and 4, 1934. RF.

Cronon, William. "Landscapes of Scarcity and Abundance." In *The Oxford History of the American West,* ed. Clyde A. Milner II et al., 603–38. New York: Oxford University Press, 1994.

Crook, George. *General George Crook: His Autobiography.* Ed. Martin F. Schmidt. Norman: University of Oklahoma Press, 1946.

Currey, George B. *Report of the Adjutant General of the State of Oregon for the Years 1865–1866.* Salem, Ore., 1866.

Darling, Ding. Letter to William Finley. May 15, 1934. RF.

————. Letter to William Finley. May 21, 1934. WFP, box 7.

Davis, Tony. "Polluted Waters Divide Oregon." *High Country News* 28 (Nov. 28, 1996).

deBuys, William, and Alex Harris. *River of Traps.* Albuquerque: University of New Mexico Press, in association with the Center for Documentary Studies at Duke University, 1990.

Deumead (?), Talbot (signature illegible; U.S. game conservation officer). Letter to George Benson. Feb. 13, 1930. RF.

Dewey, John. "Liberalism and Social Action." In *Pragmatism and American Culture*, ed. Gail Kennedy, 94–111. Boston: D. C. Heath and Co., 1950.

———. "Pragmatic America." In *Pragmatism and American Culture*, ed. Gail Kennedy, 57–60. Boston: D. C. Heath and Co., 1950.

Donegan, James. Untitled typed manuscript in Harney County Library, "Refuge" file, Burns, Ore.

Douglas, Marjory Stoneman. *The Everglades: River of Grass*. New York: Rinehart, 1947.

Duebbert, Harold F. "The Ecology of Malheur Lake and Management Implications." Refuge Leaflet 412. USDI Fish and Wildlife Service, Bureau of Sport Fisheries and Wildlife, Nov. 1969. RF.

Dunham, J. B., G. L. Vinyard, and B. E. Reiman. "Habitat Fragmentation and Extinction Risk of Lahontan cutthroat trout." *North American Journal of Fisheries Management* 17 (1997): 910–17.

Dunlap, Thomas. *Saving America's Wildlife*. Princeton: Princeton University Press, 1988.

Dunlap, R. W. (USDA acting secretary). Letter to William Finley. July 15, 1930. RF.

Dunn, Town of, Wisconsin, Park Commission. "Bike Routes." Brochure. 1983.

Durbin, Kathie. "Ranchers Arrested at Wildlife Refuge." *High Country News* 26 (Nov. 3, 1994).

———. "Restoring a Refuge: Cows Depart, but Can Antelope Recover?" *High Country News* 29 (Nov. 24, 1997).

East Oregon Herald. Burns, Ore, 1889. In Harney County Library, Burns, Ore.

Elmore, Wayne. Interview with author. June 15, 1997.

Elmore, Wayne, and J. Boone Kauffman. "Riparian and Watershed Systems: Degradation and Restoration." In *Ecological Implications of Livestock Herbivory in the West*, ed. M. Vavra et al., 212–31. Denver: Society for Range Management, 1994.

Elmore, Wayne, and Robert L. Beschta. "Riparian Areas: Perceptions in Management." *Rangelands* 9 (1987): 260–65.

Ferguson, Denzel, and Nancy Ferguson. *Sacred Cows at the Public Trough*. Bend, Ore.: Maverick Publications, 1983.

Fiege, Mark. "Creating a Hybrid Landscape: Irrigated Agriculture in Idaho." *Illahee: Journal for the Northwest Environment* 11 (1995): 60–76.

———. *Irrigated Eden: The Making of an Agricultural Landscape in the American West*. Seattle: University of Washington Press, 1999.

Findlay, John D. Letter to Col. Robert Giesen. July 28, 1969. In USACE, "Silvies River and Tributaries, Oregon: Feasibility Report for Water Resources Development" (Feb. 1977), appendix 2.

———. Letter to district engineer. Dec. 22, 1970. In USACE, "Silvies River and Tributaries, Oregon: Feasibility Report for Water Resources Development" (Feb. 1977), appendix 2.

Finley, William. "Habits and History of the Beavers." WFP, box 2/8: 1930s.

———. Letter to Ding Darling. June 2, 1934. WFP, box 7.

————. Letter to Gov. Julius E. Meier, Ore. July 2, 1934. WFP, box 7.

————. Letter to Harold C. Ickes. June 30, 1934. WFP, box 7.

————. Letter to H. M. Worcester. Sept. 27, 1934. WFP, box 7.

————. Letter to Mr. McNary. Undated. WFP, box 8, "Interior Department—Reclamation Service 1920s–1930s."

————. Letter to *Portland Oregonian.* July 28, 1930. WFP, box 8, "Interior Department—Reclamation Service 1920s–1930s."

————. "Malheur Lake." Undated, ca. 1935–38. WFP, box 9, "Klamath/Malheur Materials, 1930s."

————. "Malheur Lake–Blitzen Valley Migratory Waterfowl Refuge." Talk given Sept. 5, 1935, probably based on Stanley Jewett and William Finley, "Malheur Lake—Blitzen Valley Migratory Waterfowl Refuge," script for radio talk given August 1935. RF.

————. "Malheur Refuge an Aid to Both Birds and Farmers." Ca. 1935. WFP, box 9, Klamath/Malheur Materials, 1930s.

————. "Malheur, The Unfortunate." Ca. 1935–40. WFP, box 9.

————. "The Marshes of the Malheur." *Nature Magazine,* April 1923. RF.

————. "On the Trail of the Plume Hunters." WFP, box 7.

————. "Reclamation in the Klamath District and Its Effect on Water Fowl." WFP, box 7, Klamath-Reclamation 1934–35 folders.

————. "Reclamation vs. Conservation." *Nature Magazine* 26 (July 1935): 46–48.

————. "Trapping and Transplanting Beaver." WFP, box 2/8, 1930s.

Finley, William, and H. T. Bohlman. "Is Malheur Lake Not Worth Saving?" *Recreation Magazine,* Nov. 1916. RF.

Fitzgerald, Maurice. Testimony in "In the Matter of the Determination of the Relative Rights to the Use of the Waters of Donner und Blitzen River a Tributary of Malheur Lake." May 16, 1931. In vol. 7, "Original Evidence." Harney County Courthouse, Burns, Ore.

Fletcher, Herbert C., and Harold B. Elmendorf, "Phreatophytes—A Serious Problem in the West." In *Yearbook of Agriculture 1955: Water.* Washington, D.C.: Department of Agriculture, 1955.

Frémont, John Charles. "Report of the Exploring Expedition to the Rocky Mountains in the Year 1842, and to Oregon and North California in the Years 1843–44." Washington, D.C.: Gales and Seaton, Printers, 1845.

French, Giles. *Cattle Country of Peter French.* Portland, Ore.: Binfords and Mort, 1964.

Friedman, Ralph. *The Other Side of Oregon.* Caldwell, Idaho: Caxton Printers, 1993.

Fuller, Robert. "American Pragmatism Reconsidered: William James' Ecological Ethic." *Environmental Ethics* 14 (1992).

Gabrielson, Ira. *Wildlife Refuges.* New York: Macmillan, 1943.

Geren, Harold E. Letter to Col. Richard M. Connell. May 21, 1971. In USACE, "Silvies River and Tributaries, Oregon: Feasibility Report for Water Resources Development" (Feb. 1977), appendix 2.

Gordon, Clarence. "Report on Cattle, Sheep, and Swine, Supplementary to Enumeration of Live Stock on Farms in 1880." In *United States Tenth Census, Report on the Productions of Agriculture in the United States, 1880.* Washington, D.C.: Government Printing Office, 1883.

Gottlieb, Robert. *Forcing the Spring: The Tranformation of the American Environmental Movement.* Washington, D.C.: Island Press, 1993.

Gregory, Stanley V., Frederick J. Swanson, W. Arthur McKee, and Kenneth W. Cummins. "An Ecosystem Perspective of Riparian Zones: Focus on Links between Land and Water." BioScience 41 (1991): 540–51.

Griffin, Dorsey. *Starting at the Narrows: A History of Southern Harney County, Oregon.* Netarts, Ore.: Griffin Press, 1990.

Griffiths, David. *Forage Conditions on the Northern Border of the Great Basin, being a report upon investigations made during July and August, 1901, in the region between Winnemucca, Nevada, and Ontario, Oregon.* U.S. Department of Agriculture, Bureau of Plant Industry, bulletin no. 15. Washington, D.C.: Government Printing Office, 1902.

Guignon, Charles B., and David R. Hiley. "Biting the Bullet: Rorty on Private and Public Morality." In *Reading Rorty: Critical Responses to Philosophy and the Mirror of Nature (and Beyond),* ed. Alan R. Malachowski, 339–40. Oxford: Basil Blackwell, 1990.

Haggerty, Ancer L. "Opinion and Order." *Oregon Natural Desert Association* v. *Bureau of Land Management,* civil no. 95–2013–HA, U.S. District Court, Portland, Ore., Jan. 31, 1997.

Hanley, William. "Opinions of People Who Know the Facts About the Roosevelt Bird Bill." Ca. 1919. Hanley Company Papers. Oregon Historical Society ms. 378.

Harney Basin Development Company. "Brief in Support of the Claims of Harney Basin Development Company, before the State Water Board of Oregon. Put Forth as Evidence 26 Jan. 1918." Harney County Courthouse, Burns, Ore.

Harney Valley Items. 1887–1903. In Harney County Library, Burns, Ore.

Hawley, W. R., and C. B. McConnell, attorneys for claimant. "Contest No. 230: Against the Silvies River Irrigation Company, an Oregon Corporation; Organized for Purpose of Succeeding to Certain Rights Claimed to Have Been Theretofore Initiated by the Harney Valley Improvement Company, for Purpose of Constructing Irrigation System for 100 to 150 Thousand Acres of the Irrigable, Agricultural Sage Brush Land in the Harney Valley, by Use of Surplus Flood Waters of Silvies River and Main Upper Branch, Foley Slough." October 1923. Harney County Courthouse, Burns, Ore.

Hays, Samuel P. *Conservation and the Gospel of Efficiency: The Progressive Conservation Movement, 1890–1920.* Cambridge, Mass.: Harvard University Press, 1959.

Henshaw, F. F., and H. J. Dean. *Surface Water Supply of Oregon 1878–1910.* Water-Supply Paper 370. Washington, D.C.: Government Printing Office, 1915.

Hibbard, George. Harney County Oral History Project, oral history tape 191. 1975. Burns Library files.

Hickman, Larry, ed. *Reading Dewey: Interpretations for a Postmodern Generation.* Bloomington: Indiana University Press, 1998.

Hosford, W. E., and S. P. Pribyl. "Blitzen River Redband Trout Evaluation." Information report 83–9, Fish Division, Oregon Department of Fish and Wildlife, Portland, Ore., 1983.

Howell, P. "Oregon Chapter Workshop Raises Visibility of Declining Redbands." *Fisheries* 22 (1997): 34–36.

Hundley, Norris. *The Great Thirst: Californians and Water, 1770s–1990s.* Berkeley: University of California Press, 1992.

Hunter, Christopher J. *Better Trout Habitat: A Guide to Stream Restoration and Management.* Washington, D.C.: Island Press, 1991.

Ivey, Gary. "History of Riparian Habitat in the Blitzen Valley." Unpublished report, 1986. RF.

Ivey, Gary, John E. Cornely, and Bradley D. Ehlers. "Carp Impacts on Waterfowl at Malheur National Wildlife Refuge, Oregon." In *Transactions of the 63rd North American Wildlife and Natural Resources Conference,* 66–74. Washington, D.C.: Wildlife Management Institute, 1998.

Jackman, E. R., and John Scharff. *Steens Mountain in Oregon's High Desert Country.* Photography by Charles Conkling. Caldwell, Idaho: Caxton Printers, 1967.

Jacoby, Karl. *Crimes against Nature: Squatters, Poachers, Thieves, and the Hidden History of American Conservation.* Berkeley: University of California Press, 2001.

James, William. "On a Certain Blindness in Human Beings." In idem, *Talks to Teachers.* New York: W. W. Norton, 1958.

———. "Pragmatism." In *The Works of William James,* gen. ed. Frederick H. Burkhardt. Cambridge, Mass.: Harvard University Press, 1975.

———. "Pragmatism: A New Name for Some Old Ways of Thinking" (1907). In *The American Pragmatists: Selected Writings,* eds. Milton R. Konvitz and Gail Kennedy, 44–61. Cleveland: Meridian Books, 1960.

Jessup, L. T. (associate drainage engineer). "Report on Proposed Reflooding of a Portion of Lower Klamath Lake California." October 1927. U.S. Department of Agriculture Bureau of Biological Survey. WFP, box 7, Klamath-Reclamation 1934–35 folder.

Jewett, Stanley. Memo to Ding Darling, "Brief Report On My Activities And Of Wildlife Conditions on the Malheur Bird Refuge." July 19, 1935. RF.

———. "Report of Malheur Lake Bird Reservation June 10 to July 14, 1922." RF.

Jewett, Stanley, and Chester A. Leichhardt. Letter to Randolph S. Collins. Sept. 15, 1930. RF.

Jewett, Stanley, and William Finley. "Malheur Lake—Blitzen Valley Migratory Waterfowl Refuge." Script for radio talk given August 1935. WFP, box 7.

"John Dewey." In *The Oxford Companion to Philosophy.* Oxford: Oxford University Press, 1995. <http://w2.xrefer.com/entry/551811>.

Jordan, Terry. *North American Cattle-Ranching Frontiers: Origins, Diffusion, and Differentiation.* Albuquerque: University of New Mexico Press, 1993.

Judd, Richard. *Common Lands, Common People: The Origins of Conservation in Northern New England.* Cambridge, Mass.: Harvard University Press, 1997.

Kerr, Andy. "Steens Mountain Wilderness Act of 2000: statement on H.R. 4828, 106th Congress, 2d Session, Steens Mountain Wilderness Act of 2000, before the National Parks and Public Lands Subcommittee of the Resources Committee, U.S. House of Representatives, July 18, 2000."

Klingle, Matthew. "Urban by Nature: An Environmental History of Seattle, 1880–1970." Ph.D. diss., University of Washington, 2001.

Konvitz, Milton R., and Gail Kennedy, eds. *The American Pragmatists: Selected Writings.* Cleveland: Meridian Books, 1960.

Kostow, K. "Biennial Report on the Status of Wild Fish in Oregon." Oregon Department of Fish and Wildlife, Portland, Ore., December 1995.

Langston, Nancy. *Forest Dreams, Forest Nightmares: The Paradox of Old Growth in the Inland West.* Seattle: University of Washington Press, 1995.

Lee, D.C., J.R. Sedell, B.E. Reiman, R.F. Thurow, and J.E. Williams. "Broadscale Assessment of Aquatic Species and Habitats." In "An Assessment of Ecosystem Components in the Interior Columbia Basin and portions of the Klamath and Great Basins," eds. T.M. Quigley and S.J. Arbelbide. U.S. Forest Service General Technical Report PNW-GTR-405, pp. 1057–496. Portland, Ore., 1997.

Lewis, John H. *Irrigation in Oregon.* USDA Office of Experiment Stations Bulletin 209. Washington D.C.: Government Printing Office, 1909.

Lewis, L. Alva. "Report on Conditions of Lake Malheur Reservation, Oregon." Report for T.S. Palmer, assistant chief of USDA Bureau of Biological Survey, 1912. RF.

Li, Hiram W., and G.A. Lambaerti. "Cumulative Effects of Riparian Disturbances Along High Desert Trout Streams of the John Day Basin, OR." *Transactions of the American Fisheries Society* 123 (1994): 627–40.

Light, Andrew, and Eric Katz, eds. *Environmental Pragmatism.* London: Routledge, 1996.

Liljiqvist, L.L. (attorney for the State of Oregon). Statement to G.G. Brown, clerk of the State Land Board, April 16, 1934. WFP, box 7, Advisory Board Biological Survey.

Lo Picollo, Margaret. "Some Aspects of the Range Cattle Industry of Harney County, Oregon, 1870–1900." Master's thesis, University of Oregon, 1962.

Malanson, George P. *Riparian landscapes.* Cambridge, U.K.: Cambridge University Press, 1993.

Mandaville, Cristin R. "A Swamp in the Desert: Theory, Water Policy, and Malheur Lake Basin." M.S. thesis, Portland State University, 1996.

Markey, Merle, and Fritz Cramer. "Beaver Survey." Oregon State University Archives, RG 190, SG 2, Series 11. Project 19G box 8/1/7/60 (July–August 1937).

Martin, Anne. "Nevada: Beautiful Desert of Buried Hopes." *Nation* 115 (1926): 89–92.

Mathewson, Worth. *William Finley: Pioneer Wildlife Photographer.* Corvallis: Oregon State University Press, 1986.

Mazzoni, Joseph. *Annual Narrative 1975.* RF.

———. *Annual Narrative 1977.* RF.

Meacham, A.B. "Notes on Snakes, Paiutes, Nez Perces at Malheur Reservation." In Gordon L. Grosscup, *Paiute Indians IV,* 304. New York: Garland Publishing, 1974.

Meinzer, O.E. "Plants as Indicators of Ground Water." U.S. Geological Survey Water Supply Paper 577 (1927).

Miller, Henry. Letters to H.N. Fulgham (ranch manager). 1891. Knight Library Special Collections, University of Oregon.

Milner II, Clyde A., Carol A. O'Connor, and Martha A. Sandweiss, eds. *The Oxford History of the American West.* New York: Oxford University Press, 1994.

Mitman, Gregg. *Reel Nature: America's Romance with Wildlife on Film.* Cambridge, Mass.: Harvard University Press, 1999.

Monroe, Anne Shannon. *Feelin' Fine!: Bill Hanley's Book.* Garden City, N.Y.: Doubleday, Doran & Company, 1930.

Moran, Nathan. "Memo: Malheur Harney Lake Area." Feb. 19, 1934. RF.

Naiman, Robert J., ed. *Watershed Management: Balancing Sustainability and Environmental Change.* New York: Springer-Verlag, 1992.

Naiman, Robert J., Carol A. Johnson, and James Kelley. "Alteration of North American Streams by Beaver." *Bioscience* 38 (1988): 753–61.

Naiman, Robert J., H. DeCamps, and M. Pollock. "The Role of Riparian Corridors in Maintaining Regional Biodiversity." *Ecological Applications* 3 (1993): 209–12.

Naiman, Robert J., J.M. Melillo, and J.E. Hobbie. "Ecosystem Alteration of Boreal Forest Streams by Beaver *(Castor canadensis)*." *Ecology* 67 (1986): 1254–69.

Ogden, Peter Skene. *Snake Country Journals, 1824–25 and 1825–26.* Vol. 13. Ed. E.E. Rich. London: Hudson's Bay Record Society, 1950.

———. *Snake Country Journals, 1826–27.* Vol. 23. Ed. K.G. Davies, assisted by A.M. Johnson. London: Hudson's Bay Record Society, 1961.

———. *Snake Country Journals 1827–28 and 1828–29.* Vol. 28. Ed. Glyndwr Williams. London: Hudson's Bay Record Society 1971.

Ohmart, Robert D. "Historical and Present Impacts of Livestock Grazing on Fish and Wildlife Resources in Western Riparian Habitats." In P.R. Krausman, ed., *Rangeland Wildlife.* Denver: Society for Range Management, 1996.

Oliphant, J. Orin. *On the Cattle Ranges of the Oregon Country.* Seattle: University of Washington Press, 1968.

Oppenheimer, Todd. "The Rancher Subsidy." *Atlantic Monthly* 277 (January 1996): 26–38.

Oregon, State of. "Report of the Desert Land Board Relative to the Reclamation of Desert

Lands Granted to the State Under the Provisions of the Carey Act." 1st biennial report. Salem, Ore.: State Printer, 1911.

———. "Report of the Desert Land Board Relative to the Reclamation of Desert Lands Granted to the State Under the Provisions of the Carey Act." 2nd biennial report. Salem, Ore.: State Printer, 1913.

———. "Report of the Desert Land Board Relative to the Reclamation of Desert Lands Granted to the State Under the Provisions of the Carey Act." 5th biennial report. Salem, Ore.: State Printer, 1921.

———. "Report of the Desert Land Board Relative to the Reclamation of Desert Lands Granted to the State Under the Provisions of the Carey Act." 6th biennial report. Salem, Ore.: State Printer, 1923.

Oregon Daily. Feb. 21, 1935. RF.

Oregon Department of Fish and Wildlife. "Work Must Continue to Protect Oregon's Redband Trout." Press release, March 23 2000.

Oregon Natural Desert Association. "Environmental Groups Disappointed that Unique Desert Fish Will Receive No Protection." Press Release, March 20, 2000.

Oregon Natural Desert Association, Oregon Trout, Native Fish Society, and Oregon Council of Trout Unlimited. "A Petition for Rules to List Great Basin Redband Trout *(Oncorhynschus mykiss ssp)* as Threatened or Endangered Under the Endangered Species Act." Sept. 4, 1997.

Ortega, Prim. "Testimony in the Matter of the Determination of the Relative Rights to the Use of the Waters of Donner und Blitzen River, a Tributary of Malheur Lake." May 16, 1931. In vol. 7, "Original Evidence," 316–68. Harney County Courthouse, Clerk's Office, Burns, Ore.

Otley, Fred. Letter to E. W. Nelson. June 14, 1922. RF.

———. "Malheur Lake Riparian Owners Plea." June 1922. Malheur Lake Riparian Owners Association. RF.

Otley, Fred (private rancher, descendent of Fred Otley above). Interview with author, June 1977.

Pacific Live Stock Company. "In the Supreme Court of the State of Oregon, Pendleton term May 1924, In the Matter of the Determination of the Relative Rights to the Waters of the Silvies River and its Tributaries in Grant and Harney Counties, Oregon." Appeal from the Circuit Court of the State of Oregon, Appellant's Abstract of Record, Pacific Livestock Company Appellant. Harney County Courthouse, Burns, Ore.

Parker, Kelly A. "Pragmatism and Environmental Thought." In *Environmental Pragmatism,* eds. Andrew Light and Eric Katz, 21–37. London: Routledge, 1996.

Pearson, T. Gilbert. Letter to Frank Triska. 1916. RF.

Pisani, Donald J. "Beyond the Hundredth Meridian: Nationalizing the History of Water in the United States." *Environmental History* 5 (2000): 466–82.

————. *To Reclaim a Divided West: Water, Law, and Public Policy, 1848–1902*. Albuquerque: University of New Mexico Press, 1992.

————. *Water, Land, and Law in the West: The Limits of Public Policy, 1850–1920*. Lawrence: University Press of Kansas, 1997.

Portland Oregonian. "Antelope Refuge Will Look at Goals." Feb. 1, 1998.

————. Feb. 23, 1935. RF.

————. "The Fine Print on the Steens." Editorial. July 18, 2000.

————. "The Steens Legislation." July 19, 2000.

Price, Jennifer. "When Women Were Women, Men Were Men, and Birds Were Hats." In idem, *Flight Maps: Adventures with Nature in Modern America*, 57–110. New York: Basic Books, 1999.

Prince, High. *Wetlands of the American Midwest: A Historical Geography of Changing Attitudes*. Chicago: University of Chicago Press, 1997.

Raleigh, Robert (superintendent of Squaw Butte Experiment Station). Letter to Col. Allaire. Dec. 14, 1976. In USACE, "Silvies River and Tributaries, Oregon: Feasibility Report for Water Resources Development" (Feb. 1977), appendix 2.

Reed, Nathaniel P., and Dennis Drabelle. *The United States Fish and Wildlife Service*. Boulder, Colo.: Westview Press, 1984.

Rinehart, William. "Report of the Commission of Indian Affairs, 1876." Reprinted in A. B. Meacham, "Notes on Snakes, Paiutes, Nez Perces at Malheur Reservation," in Gordon L. Grosscup, *Paiute Indians IV*, 271–73. New York: Garland Publishing, 1974.

————. "Report of the Commission of Indian Affairs, 1878." Reprinted in A. B. Meacham. "Notes on Snakes, Paiutes, Nez Perces at Malheur Reservation," in Gordon L. Grosscup, *Paiute Indians IV*, 281–88. New York: Garland Publishing, 1974.

————. "Report of the Commission of Indian Affairs, 1879–1880." Reprinted in A. B. Meacham, "Notes on Snakes, Paiutes, Nez Perces at Malheur Reservation," in Gordon L. Grosscup, *Paiute Indians IV*, 289–306. New York: Garland Publishing, 1974.

Robbins, William. *Landscapes of Promise: The Oregon Story 1800–1940*. Seattle: University of Washington Press, 1997.

Robinson, Michael. *Water for the West: the Bureau of Reclamation, 1902–1977*. Chicago: Public Works Historical Society, 1979.

Robinson, T. W. "Phreatophytes." U. S. Geological Survey Water Supply Paper 1423 (1959): 1–84.

Rorty, Richard. *Consequences of Pragmatism*. Minneapolis: University of Minnesota Press, 1982.

————. "On Ethnocentrism: A Reply to Clifford Geertz." In idem, *Objectivity, Relativism and Truth: Philosophical Papers*, vol. 1.: 203–10. Cambridge, U.K.: Cambridge University Press, 1991.

————. "The Priority of Democracy to Philosophy." In idem, *Objectivity, Relativism and*

Truth: Philosophical Papers, vol. 1: 175–96. Cambridge, U.K.: Cambridge University Press, 1991.

Rosenthal, Sandra B., and Rogene A Buchholz. "How Pragmatism *Is* an Environmental Ethic." In *Environmental Pragmatism,* ed. Andrew Light and Eric Katz, 38–49. London: Routledge, 1996.

Roy, Richard. "Blitzen River Fish Habitat Improvement Strategy." No date (after 2000). RF.

——— (supervisory wildlife biologist, Malheur National Wildlife Refuge). Interview with author, Jan. 10–11, 2002.

Russell, Israel C. *Notes on the Geology of Southwestern Idaho and Southeastern Oregon.* U.S. Department of the Interior, United Stated Geological Survey, bulletin no. 217. Washington, D.C.: Government Printing Office, 1903.

Scharff, John C. "A Brief History of Malheur Refuge, Malheur Lake Area—A Scene of a Thousand Interests." 1938. RF.

———. "Historical Material—P-Ranch." 1972. RF.

———. "Report of Activities." Fiscal years 1937, 1938, 1939, 1940. Malheur Migratory Waterfowl Refuge. RF.

Scott, James C. *Seeing Like a State.* New Haven: Yale University Press, 1998.

Sharp, Dallas Lore. *Where Rolls the Oregon.* Boston: Houghton Mifflin Co.: Riverside Press, 1914.

Sheldon, H. Letter to George Benson. Nov. 8, 1929. RF.

Shirk, David L. *The cattle drives of David Shirk from Texas to the Idaho mines, 1871 and 1873: reminiscences of David L. Shirk, wherein are described his two successful cattle drives from Texas, in company with George T. Miller. His later experiences as a cattleman in eastern Oregon during the terrible depredations of hostile Indians and the range warfare with Pete French, from the original manuscript and related papers, now in the University of Oregon Library.* Ed. Martin F. Schmitt. Portland, Ore.: Champoeg Press, 1956.

Shutkin, William. *The Land That Could Be: Environmentalism and Democracy in the Twenty-First Century.* Cambridge, Mass.: MIT Press, 2000.

Simpson, Peter K. *The Community of Cattlemen: A Social History of the Cattle Industry in Southeastern Oregon 1869–1912.* Moscow, Idaho: University of Idaho Press, 1987.

Sokol, Chris. "Rivers from a Timber Industry Perspective." In *Riparian Management: Common Threads and Shared Interests.* In "General Technical Report—Rocky Mountain Forest and Range Experiment Station, Fort Collins" ("GTR—RM"), 1993: 37–38.

Southworth, Jack. "Ranching and Riparians." Talk given in May 1993 at the Blue Mountains Natural Resources Institute. Text reprinted in *Natural Resource News,* August 1993: 12–13.

Steinberg, Theodore. *Slide Mountain: Or the Folly of Owning Nature.* Berkeley: University of California Press, 1995.

Steward, Julian H., and Erminie Wheeler-Voegelin. *The Northern Paiute Indians.* New York: Garland, 1974.

Stivers, H. R. (acting regional director of the Bureau of Reclamation). Letter to Col. C. J. Allaire (district engineer, Walla Walla). Ca. Dec. 14, 1976–Jan. 26, 1977. In USACE, "Silvies River and Tributaries, Oregon: Feasibility Report for Water Resources Development" (Feb. 1977), appendix 2.

Stone, H. F. Letter to George Benson. July 20, 1920. RF.

Stuebner, Stephen. "Go Tell It on the Mountain." *High Country News* 31 (Nov. 22, 1999).

Talbot, Caryn. "P Ranch: History, Preservation and Interpretive Development." 1976. RF.

Talmer, T. S. Letter to George Cantwell. Jan. 18, 1916. RF.

Thomas, Jack Ward, Chris Maser, and J. E. Rodiek. "Wildlife Habitats in Managed Rangelands—The Great Basin of Southeastern Oregon: Riparian Zones." USDA Forest Service General Technical Report PNW-80. 1979.

Thompson, Paul B. "Pragmatism and Policy: The Case of Water." In *Environmental Pragmatism,* ed. Andrew Light and Eric Katz, 187–208. London: Routledge, 1996.

Treadwell, Edward, and John L. Rand. "Brief in Support of Claims of Pacific Live Stock Company." In *In the Matter of the Determination of the Relative Rights to the Waters of Silvies River and Its Tributaries, a Tributary of Malheur Lake. In the Circuit Court of the State of Oregon for Harney County.* Testimony taken 1918; findings dated October 19, 1923. Harney County Courthouse, Burns, Ore.

U.S. Army Corps of Engineers, Walla Walla District (USACE). "Final Version, Malheur Lake Flood Damage Reduction Feasibility Study and Environmental Impact Statement." April 1987.

———. "Reconnaissance Report, Malheur Lake, Oregon." August 29,1985.

———. "Silvies River and Tributaries, Oregon: Feasibility Report for Water Resources Development." Feb. 1977.

U.S. Army Engineer District, Corps of Engineers. "Survey Report on Silvies River and Tributaries, Oregon." Portland, Nov. 8, 1957.

USDA Economic Research Service. "Report on Water and Related Land Resources Malheur Lake Drainage Basin Oregon. " Based on a cooperative survey by the State Water Resources Board of Oregon and the USDA. April 1967.

USDA Forest Service. *Malheur National Forest Land and Resource Management Plan, Final Environmental Impact Statement.* Washington: Government Printing Office, 1990.

USDA/USDI. "Interior Columbia Basin Ecosystem Management Project, Eastside Draft Environmental Impact Statement." Vol. 1. Walla Walla, Wash., May 1997.

U.S. Department of Commerce, Bureau of the Census. *Tenth Census of the United States: 1880, Report on Agriculture.* Washington, D.C.: Government Printing Office, 1883.

U.S. Department of the Interior (USDI). *Report of the Secretary of the Interior 1885–88.* House Executive Documents, U.S. Congress, 49th Congress, 1st Session, 198 (series 2378).

USDI Bureau of Indian Affairs. "The Burns Paiute Colony: Its History, Population and Economy." Report 227, prepared by the Planning Support Group in Cooperation with the Burns-Paiute Colony, 1974. On file in Burns Paiute tribal office, Burns, Ore.

USDI Bureau of Land Management. "South Steens Allotment Management Plan and Environmental Assessment." Burns District Office, Hines, Ore., June 1995.

———. "Water Quality Management Plan: South Fork Donner und Blitzen River, Home Creek, Skull Creek, Threemile Creek." Burns District Office, Hines, Ore., June 22, 1998.

USDI Bureau of Reclamation. Data on Klamath Project. <http://dataweb.usbr.gov/ html/ klamath.html#plan>.

USDI Fish and Wildlife Service. "Compromise Committee Report." RF.

———. "Notice of Petition Finding." 50 CFR, part 17, *Federal Register,* vol. 65, no. 54 (March 20, 2000): 14932–36.

———. Bureau of Sport Fisheries and Wildlife. "Carp Control Project at Malheur Lake, Oregon 1955–1956." 1957. RF.

———. Bureau of Sport Fisheries and Wildlife. *Master plan, Malheur National Wildlife Refuge.* Portland, Ore., 1965.

———. Division of Ecological Services, "Planning Aid Letter on the Malheur Lake Flood Control Project." Portland, April 10, 1985. In USACE, "Final version, Malheur Lake Flood Damage Reduction Feasibility Study and Environmental Impact Statement" (April 1987), appendix C.

———. "Draft Environmental Impact Statement for the Hart Mountain National Antelope Refuge Comprehensive Management Plan." Portland, Ore.: U.S. Fish and Wildlife Service Region 1, July 1993.

———. Malheur National Wildlife Refuge. *Annual Narrative* (various years). RF.

———. Malheur National Wildlife Refuge. *Blitzen Valley Management Plan 1941.* RF.

———. Malheur National Wildlife Refuge. *Quarterly Narrative Report* (various months and years). RF.

U.S. General Accounting Office. *Public Rangelands: Some Riparian Areas Restored But Widespread Improvement Will be Slow: Report to Congressional Requesters.* Washington, D.C.: Government Printing Office, 1988.

Vander Shaff, Dick. "Final Report: Donner Und Blitzen Wild and Scenic River Sensitive Plants and Unique Natural Areas Inventory." 1992. Unpublished Nature Conservancy report, on file in Oregon Natural Desert Association Office, Portland, Ore.

Vavra, M., W. A. Laycok, and R. D. Pieper (eds), *Ecological Implications of Livestock Herbivory in the West.* Denver: Society for Range Management, 1994.

Vileisis, Ann. *Discovering the Unknown Landscape: A History of America's Wetlands.* Washington, D.C.: Island Press, 1997.

Wallace, H. A. Letter to Hon. Walter M. Pierce. Feb. 27, 1935. RF.

Wallen, Henry D. "Report of Expedition." In "Report of the Secretary of War, 1859–1860." Serial 1031, Senate document, 36th Congress, 1st session, Executive Document No. 34.

Walters, Timothy Robert. *Surviving the Second Civil War: The Land Rights Battle . . . and How to Win It.* Safford, Ariz.: Rawhide Western Publishing, 1994.

Waring, Gerald A. "Geology and Water Resources of the Harney Basin Region, Oregon." USDI U.S. Geological Survey Water-Supply Paper 231. Washington, D.C.: Government Printing Office, 1909.

Warren, Louis S. *The Hunter's Game: Poachers and Conservationists in Twentieth-Century America.* New Haven: Yale University Press, 1997.

West Shore. February 1885.

Whistler, John T., and John H. Lewis. "Harney and Silver Creek Projects: Irrigation and Drainage." Oregon Cooperative Work, Department of the Interior United States Reclamation Service. Denver: Smith-Brooks Press, 1916.

White, Richard. *The Organic Machine: The Remaking of the Columbia River.* New York: Hill and Wang, 1995.

Wilde, James Dale. "Prehistoric Settlements in the Northern Great Basin: Excavations and Collections Analysis in the Steens Mountain Area, Southeastern Oregon." Ph.D. diss., University of Oregon, 1985.

Wiley, A. J. "Report 1922: Blitzen River Reclamation District, Harney County Oregon." On file in Water Masters Office, Burns, Ore.

Wiley, Ron, and Guy Sheeter. "Water Quality Parameters." In USDI Bureau of Land Management, "South Steens Allotment Management Plan and Environmental Assessment," Burns District Office, Hines, Ore. (June 1995), appendix C.

Willet, George. Reservation Inspector. "Report of Malheur Lake Bird Reservation." Submitted to E. W. Nelson, chief of the Biological Survey. Sept. 28, 1918. RF.

"William James. "In *The Oxford Companion to Philosophy.* Oxford: Oxford University Press, 1995. <http://w2.xrefer.com/entry/552470>.

Wilson, Rocky. "'Clean Stream Initiative' Closes in on Fall Election." *Wallowa County Chieftain,* Nov. 1996.

Worster, Donald. "A Country without Secrets." In idem, *Under Western Skies,* 238–54. Oxford: Oxford University Press, 1994.

———. *Rivers of Empire: Water, Aridity, and the Growth of the American West,* 31–54. New York: Pantheon Books, 1985.

———. "Water as a Tool of Empire." In idem, *An Unsettled Country: Changing Landscapes of the American West.* Albuquerque: University of New Mexico Press, 1994.

Young, James B., and B. Abbot Sparks. *Cattle in the Cold Desert.* Logan: Utah State University Press, 1990.

Young, James B., and Jerry D. Buddy. "Historical Use of Nevada's Pinyon-Juniper Woodlands." *Journal of Forest History* 23 (1979): 113–21.

INDEX

Illustrations are indicated with "pl" for plate.